Craniofacial Biology of Primates

Symposia of the Fourth International Congress of Primatology

Portland, Ore., August 15–18, 1972

Vol. 3

Main Editor: WILLIAM MONTAGNA
Oregon Regional Primate Research Center, Beaverton, Ore.

S. Karger · Basel · München · Paris · London · New York · Sydney

Craniofacial Biology of Primates

Volume Editor: M. R. ZINGESER
Oregon Regional Primate Research Center, Beaverton, Ore.

With 66 figures and 29 tables

S. Karger · Basel · München · Paris · London · New York · Sydney 1973

Symposia of the Fourth International Congress of Primatology

S. Karger · Basel · München · Paris · London · New York · Sydney
Arnold-Böcklin-Strasse 25, CH–4011 Basel (Switzerland)

Contents

Contents

Preface

The primate craniofacial complex is of interest to a remarkably broad spectrum of investigators including paleontologists, neozoologists, and clinical researchers. The Craniofacial Biology Program of the Fourth International Congress of Primatology was organized to reflect the scope of these interests in so far as was practical. Primate evolution and systematics, physiology, anatomy, biochemistry, and behavior have been the subjects of intensive research for years; consequently, there is an extensive literature. With the burgeoning of the use of nonhuman primates in research, this body of information can no longer be neglected as irrelevant by the clinically-oriented investigator. These laboratory animals are often as different from each other as they are from men. A knowledge of the differences, and conversely of the degree of kinship of these species is essential for conscientious experimental design and valid conclusions; its absence, however, is all too often regrettably apparent in much of the published work. This is not to say that the informational void is confined to the clincial researcher. A knowledge of the evolutionary process, the unifying and pervasive biological principle that should be a part of the intellectual armamentarium of all men, is obviously sadly neglected in many 'traditionally' structured biological disciplines.

The objective of interdisciplinary exchange, repeatedly enunciated in numerous introductory and prefatory statements, is here reiterated. It has served to guide the format of this full-day section of the Congress. Two formal symposia were scheduled, and papers contributed by the participants are selectively included in this volume. In addition, contributions by J. R. E. MILLS and P. M. BUTLER, who were unable to attend, appear. Most of the topics covered in the first symposium, Evolution, are presented in the related groupings of the evolution of the masticatory system (BUTLER, HIIEMÄE and KAY, and MILLS) and dental morphology and systematics (GREENE, KINZEY, KITAHARA-FRISCH, VON KOENIGSWALD, and STEELE).

The second symposium, Developmental and Functional Anatomy, was organized with the objective of elucidating primate craniofacial growth and related form-functional correlations. B. C. MOFFETT's paper synthesizes the results of a unique graduate research program at the University of Washington's Department of Orthodontics. This program uses nonhuman primates to analyze clinical mechanotherapeutic procedures, thereby contributing to our knowledge of craniofacial growth and functional dynamics. M. L. Moss documents his views on the functional matrix and related concepts, together with their analytic potentiality. B. G. SARNAT, a pioneer in the use of monkeys for craniofacial growth studies, reviews some surgical experimentation. M. R. ZINGESER presents comparative evidence quantitatively substantiating the integrated nature of the occlusofacial complex and touches upon the clinical implications of these findings. D. R. SWINDLER's comparative growth study is also included in this section of the book.

The contributors to this second symposium are mostly either clinician-scientists or investigators associated with clinical disciplines. Since anthropologists and paleontologists are also interested in growth and functional anatomy, the nature of these findings bridges the gap between the more 'esoteric' fields (as viewed from a medical standpoint) and the clinical subject matter of the third part of the Craniofacial Biology Program.

The third part of the Program, an informal discussion seminar, entitled Application to Clinical Problems of Basic Research in Craniofacial Biology, was organized and chaired by J. E. HAMMER III of the National Institute of Dental Research; the short papers issuing from this session were published in the *Proceedings* of the Congress (the March 1973 issue of the *American Journal of Physical Anthropology*). In connection with the clinical aspects of the program, KENNETH HISAOKA of the National Institute of Dental Research spoke on Nonhuman Primates as a Resource in Craniofacial Research. His remarks appear as an appendix to this volume.

A conference program inevitably reflects the interests and predilections of its chairman. I accept this judgement with special acknowledgement to the two men most influential in inducing in me any bias apparent in this volume. I am especially beholden to MILO HELLMAN, clinician-anthropologist-paleontologist, and WILLIAM KING GREGORY, whose odontologic studies are still of great relevance after half a century; these sometime collaborators were the 'models' guiding my early professional interests.

Finally, I wish to thank my co-chairman, Dr. A. A. DAHLBERG of the University of Chicago Department of Anthropology, for his valued assistance and advice and Dr. LOUIS G. TERKLA, Dean of the University of Oregon Dental School, for his efforts on behalf of our Craniofacial Biology Program. I gratefully acknowledge the cooperation given me by Drs. WILLIAM MONTAGNA and THEODORE GRAND of the Oregon Regional Primate Research Center. This volume would not be possible without the help of Mrs. ROBERT LOW, the Congress Coordinator and *ad hoc* copy editor; her hard work and that of the other staff members of the Center are sincerely appreciated.

Portland, Oregon   M. R. ZINGESER
October, 1973

The Evolution of the Masticatory System

Symp. IVth Int. Congr. Primat., vol. 3: Craniofacial Biology of Primates, pp. 1–27 (Karger, Basel 1973)

Molar Wear Facets of Early Tertiary North American Primates

P. M. BUTLER

Royal Holloway College, University of London, Englefield Green, Surrey

Introduction

The wear facets on fossil molar teeth are of interest because they provide information about occlusal relations and the relative movements of the opposing teeth during chewing. They enable us to relate structure to function and thus to get an inkling of the adaptive significance of evolutionary change in the dentition. However, although the older literature contains many incidental references to wear facets, their systematic study in a group of fossil mammals was first made only in 1952, when it was shown that throughout the Perissodactyla the facets could be homologized [BUTLER, 1952a, 1952b]. MILLS [1955, 1963, 1966] found facets corresponding to those of perissodactyls in recent primates and insectivores. Comparison of divergently specialized derivatives of the tribosphenic molar pattern showed that intercuspal relations are retained by correlative evolution of upper and lower teeth, so that the evolutionary history of each facet can be traced [BUTLER, 1961]. The study of wear facets has been extended to Mesozoic mammals with pretribosphenic molar patterns [MILLS, 1964; CROMPTON and JENKINS, 1968].

From scratches and grooves on the worn surfaces of *Hyracotherium* molars it was concluded [BUTLER, 1952a] that movement of the lower teeth while in contact with the upper teeth had an appreciable transverse (ectental) component, amounting to about half the width of the upper molar. At the same time, there was a vertical (orthal) component which varied within each chewing cycle: when the buccal cusps of both jaws were in contact, the jaws were closing; but as the lower teeth passed more lingually, their upward movement was reduced and eventually reversed. MILLS

[1955] distinguished in primates between (a) an ectental closing movement centered near the condyle on the same side as the teeth (ipsilateral), which he called the buccal phase of occlusion, and (b) an oblique and more horizontal movement centered near the opposite (contralateral) condyle, constituting the lingual phase. The buccal and lingual phases coincided at the centric position, and in most primates buccal-phase movement on one side of the mouth took place at the same time as lingual-phase movement on the other. A distinct set of facets is produced during the lingual phase that can be distinguished by the direction of their striations [BUTLER and MILLS, 1959].

CROMPTON and HIIEMÄE [1969, 1970] made a cineradiographic study of *Didelphis*, combined with a detailed study of its molar wear facets. They found no evidence of a lingual phase and inferred that this mode of occlusion was lacking in primitive tribosphenic mammals generally. They showed that in *Didelphis* molar function was unilateral, the teeth on the non-working side being out of contact. They drew a distinction between centric *occlusion*, when the upper and lower teeth are maximally interlocked, and the centric *relation*, when the mandibles are symmetrically placed in relation to the skull. They showed that movement at the symphysis was remarkably free, a fact which MILLS had insufficiently allowed for in his studies based on dried skulls.

During a visit to the United States in 1965/66, I took the opportunity to examine a wide range of Palaeocene and Eocene mammals, including most genera of North American fossil primates. Particular attention was paid to molar wear facets and especially to any that might have been produced in a lingual-phase movement like that postulated in recent primates by MILLS. The work of CROMPTON and HIIEMÄE prompts me to publish my findings, even though owing to lack of time I was unable to study more than a small proportion of the available specimens and the observations are less complete than I would wish.[1]

1 That such a rapid survey was indeed wide enough to produce a result at all is due very largely to the generous help provided by the many palaeontologists in the United States and Canada who gave me access to the collections under their charge and facilities to study them. I would like to take this opportunity to express my sincere thanks, especially mentioning Dr. C. L. GAZIN, at the U. S. National Museum (USNM), Dr. C. C. BLACK, at the Carnegie Museum, Dr. M. C. McKENNA and Dr. L. VAN VALEN, at the American Museum of Natural History (AMNH), Dr. P. ROBINSON, at the University of Colorado, Prof. G. L. JEPSEN, at Princeton, Dr. P. O. McGREW, at the University of Wyoming, Dr. W. A. CLEMENS, at the University of Kansas, and Dr. L. S. RUSSELL and Dr. G. EDMUND, at the Royal Ontario Museum.

The Nature of Wear Facets

It is proposed to restrict the term *facet* to wear that is produced when teeth are nearly or actually in contact with their opponents, when the food has been broken down sufficiently to permit upper and lower cusps to interdigitate. Where the teeth are pressed together, their relative movement, aided by hard particles in the food, produces areas of polished enamel. Such facets appear early in the life of the tooth and tend to increase in size as the wear proceeds. They are typically flat surfaces which reflect light; but very often they are scored with fine parallel striations, indicative of the relative movement of the teeth by which they were produced. To each facet on an upper tooth there is a corresponding facet on a lower tooth. Facets are most easily seen on the enamel; but as they develop, dentine may become exposed.

Facet wear may be distinguished from food abrasion, which occurs when the teeth are not in close contact.[2] Large pieces of food may be held by prominent parts of the teeth while the jaws are partly open, as in the crushing of seeds and fruits or the breaking of the carapaces of large arthropods. The tips of the cusps wear away from such a cause, resulting in a general flattening of the crown. Cusps which do not contact the opposing teeth, e.g., the buccal styles of opossums and zalambdodont insectivores, are nevertheless worn by holding and piercing the food. CROMPTON and HIIEMÄE showed that such a process is an important constituent of mastication in *Didelphis*. Besides the wear of the tips of cusps, there is a general abrasion of the tooth surface resulting in removal of enamel wrinkles, rounding of ridges, and hollowing out of areas of exposed dentine. Dentine exposed by abrasion forms a crater surrounded by a prominent ridge of enamel, rounded in section; whereas dentine exposed by facet wear is bordered by enamel that has the appearance of having been filed away to form a flat surface. Food abrasion does not as a rule result in flat areas such as characterize facet wear, and scratches on the worn surface are randomly distributed.

MacINTYRE [1966], working on Miacidae, made a similar distinction between 'shear facets' and 'non-occlusal attrition'. Within the latter category he distinguished a form of wear that he called horizontal attrition, due to chewing hard substances such as bone, which wear the apices of the cusps smoothly flat.

2   Food abrasion and facet wear as used here, correspond to abrasion and attrition as defined by STONES [1948].

DAHLBERG and KINZEY [1962] distinguished between 'abrasion', due to contact of teeth with foreign bodies, and 'attrition', due to tooth-on-tooth contact. They used replica techniques on human teeth. Abrasion produced scratches or striations that ran predominantly in two directions, corresponding to rotation of the jaw around the ipsilateral or contralateral condyle; but attrition resulted only in polishing of the tooth surface. The two types of wear do not correspond exactly with food abrasion and facet wear as defined in this paper, for scratches frequently occur on facets.

EVERY and KÜHNE [1970] attribute striated wear surfaces to a process of 'thegosis', i.e. the wear of tooth against tooth with the function of sharpening cutting edges. CROMPTON and HIIEMÄE [1970] distinguished a 'crushing-puncturing stroke' in the earlier stage of mastication from a 'shearing stroke' with tooth-to-tooth contact in the later stage. It is quite possible that one effect of the shearing stroke is to sharpen cutting edges and so increase their effectiveness in the earlier stage of subsequent mastication. Nevertheless, the shearing stroke is an integral part of the mastication process, and not a process independent of mastication, as seems to be implied in the thegosis concept [EVERY, 1965]. At least as far as the molars are concerned, there is no evidence that the movements involved in thegosis are in any way different from those of the later stage of mastication. Nevertheless, thegosis is a convenient word to describe movements whose function appears to be primarily that of tooth sharpening, as the grinding of the incisors of rabbits or the canines of pigs [EVERY, unpublished].[3]

Identification of Wear Facets

Facets are most easily seen by illuminating the tooth with a beam of light from a lamp source and rotating the tooth until the light reflected from its surface reveals a flat area. Facets that are parallel to each other will be illuminated together (fig. 1). The facets are then examined carefully, under higher magnification if necessary, for evidence of scratch lines; and the direction of these is noted (fig. 2). All striations produced by a single jaw movement will be parallel (strictly speaking, on

3   ZINGESER [1968, 1969, 1971] describes honing morphology in primate dentitions as being particularly obvious in the case of the canine teeth, but not limited to these teeth [1968]. However, special honing behavior is characteristic of males with large maxillary canine teeth; while other teeth in both sexes are sharpened in the course of the masticatory cycle. [Ed.]

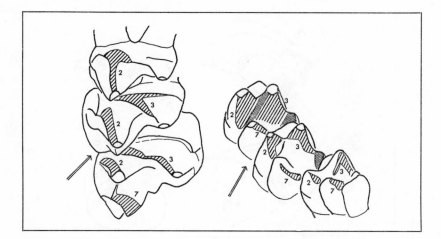

Fig. 1. Left: *Notharctus tyrannus,* anterior view of right P³–M¹, R. Ont. Mus. 3068. Right: *Pelycodus trigonodus,* posterior view of left M_1–M_3, R. Ont. Mus. 917. Shaded to show series of parallel wear facets produced by relative movement in the direction of the arrow.

concentric arcs); and it follows that where two facets, produced during the same movement meet to form a ridge or groove (e.g., the posterior facet on the paracone and the anterior facet on the metacone), the ridge or groove must be parallel to the direction of jaw movement. If the tooth is rotated so as to be seen in the direction of the striations, i.e., in the direction of relative movement of the opposing tooth, then all the facets produced by that movement will be seen tangentially, on the profile of the cusps. Conversely, by rotating the tooth till the facets are simultaneously at the point of disappearing from view, one can determine the direction of relative movement that produced them. Any facets which remain visible after this procedure has been followed must have been produced by jaw movement in a different direction. In no case was it necessary to postulate more than two directions of jaw movement to account for all the facets on a tooth.

The relations between upper and lower facets were determined by superimposing *camera lucida* drawings of the teeth. Particularly useful were drawings of teeth seen in the direction of relative motion and illustrated by figure 3. In the buccal phase of occlusion, the lower molar moves lingually and upwards across the upper molar; all the facets so produced are seen in profile by looking at the upper molar from the buccal side in an obliquely upward direction. At the same time, the relative movement of the upper molar is buccally and downward across the lower molar; and the lower molar facets are seen in profile if the tooth is looked at from the lingual side in an obliquely downward direction. If drawings of the two teeth are now superimposed, one can imagine the lower molar travelling perpendicularly to the paper. The V-shaped groove between the protoconid and the hypoconid should fit the V-shaped profile of the paracone, and the occluding facets on the arms of the

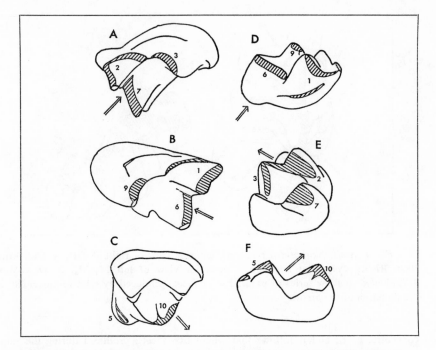

Fig. 2. Molars of *Omomys carteri,* to show wear facets. *A–C* M¹ right, Carn. Mus. 6491. *D-F* M₁ right, Carn. Mus. 6395. *A, D* Forward-facing facets of the buccal phase. *B, E* Backward-facing facets of the buccal phase. *C, F* Lingual-phase facets. Arrows show directions of relative movement in which the facets were produced.

two V's should coincide. Similarly, the hypoconid should fit the groove between the paracone and the metacone; and the protocone should fit the groove between the metaconid and the entoconid. If two adjacent teeth are included, the relations of the metacone and hypocone of the upper molar to the trigonid of the more posterior lower molar may be seen. Lingual-phase facets will still remain visible on such drawings, and their mode of formation can be easily determined if the direction of motion is known from scratches. In most cases it was impossible to obtain occluding teeth of the same individual, and some allowance had to be made for differences of wear and variation in size; however, by choice of suitable specimens it was usually possible to reduce these to a minimum. As most of the material consisted of fragments of maxilla and mandible and since skulls, where available, had in nearly every case been crushed or distorted, movements of the condyles could be reconstructed only in a very approximate manner in a small number of genera.

There is no generally recognized system of nomenclature for wear facets, and each author has used a different system. This situation is inev-

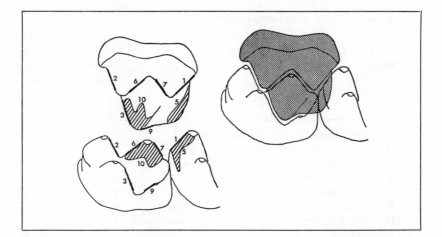

Fig. 3. Palaechthon sp. U. Kans. 9558, 9560. Left: M¹ left and M$_{1-2}$ right, seen in the direction of relative movement. All the facets except 5 and 10 (shaded) are seen in profile. Right: superposition of drawings to show buccal-phase occlusion. The upper molar is imagined as seen in transparency, and the lower teeth as moving perpendicularly to the paper.

itable so long as the present state of uncertainty remains over the evolution of mammalian molar occlusion. If, as now seems probable, molar facets can be homologized throughout the therian mammals, an agreed-upon system of naming them will have to be established. For the purposes of the present paper, the facets of primates are numbered to correspond with those of perissodactyls [BUTLER, 1952a], as given in table I.

Material Examined

A total of 122 specimens of North American Palaeocene and Eocene primates were studied, of which 49 contained upper molars and 73 had lower molars. Representatives of 21 genera in seven families were included, as shown in table II.

Although the sample from each genus was small, the group as a whole presented a remarkably uniform picture; and it seems unlikely that the inclusion of more specimens would materially alter this. In the description which follows, all the families are treated together, except the Carpolestidae.

Table I. Molar wear facets of primates [1]

Lower molar facets	Upper molar facets	*Hyracotherium* [BUTLER, 1952a]	*Didelphis* [CROMPTON and HIIEMÄE, 1970]
Protoconid, anterior	metacone, posterior	1	1
Protoconid, posterior	paracone, anterior	2	2
Metaconid, posterior	protocone, anterior	3	4
Metaconid, anterior	hypocone, posterior	4	absent
Protoconid, lingual	hypocone, buccal	5	absent
Hypoconid, anterior	paracone, posterior	6	3
Hypoconid, posterior	metacone, anterior	7	6
Entoconid, posterior	hypocone, anterior	8	absent
Entoconid, anterior	protocone, posterior	9	5
Hypoconid, lingual	protocone, buccal	10	absent

1 All these facets can be recognized in early primates, with the exception of 4 and 8. In Notharctidae, facets analogous to these two develop on the pseudohypocone and are designated 4' and 8'.

Table II. Specimens included in this study

Microsyopidae	*Microsyops* (L.-M. Eocene)
Plesiadapidae	*Pronothodectes* (M. Palaeocene), *Plesiadapis* (U. Palaeocene)
Carpolestidae	*Elphidotarsius*[1] (M. Palaeocene), *Carpodaptes* (U. Palaeocene), *Carpolestes* (U. Palaeocene)
Notharctidae	*Pelycodus* (L. Eocene), *Notharctus* (M. Eocene), *Smilodectes* (M. Eocene)
Paromomyidae	*Palaechthon* (M. Palaeocene), *Paromomys* (M. Palaeocene), *Phenacolemur* (U. Palaeocene – L. Eocene)
Anaptomorphidae	*Palenochtha* (M. Palaeocene), *Absarokius*[1] (L. Eocene), *Tetonius* (L. Eocene), *Anaptomorphus*[1] (M. Eocene), *Uintanius*[1] (M. Eocene)
Omomyidae	*Shoshonius*[1] (L. Eocene), *Omomys* (M. Eocene), *Hemiacodon* (M. Eocene), *Washakius* (M. Eocene)

1 Genera represented by lower teeth only.

Wear Facets on the Molars

The wear facets described in this section refer to those produced by interactions between M_1 and M^1, M^1 and M_2, M_2 and M^2, and M_2 and M^3. Those between M_3 and M^3 and between P^4 and M_1 are described in later sections. The cusps and ridges are named in accordance with the scheme proposed by VAN VALEN [1966].

A. Buccal-Phase Facets (fig. 1, 2, 3, 4, 7)

Facet 2 occupies the edge of the anterior crest of the paracone (paracrista), the parastyle, the adjacent part of the anterior cingulum as far lingually as the paraconule (paracingulum), and the anterior surface of the paraconule. It faces anteriorly, downwards, and somewhat lingually. The corresponding facet on the lower molar occupies the posterior surface of the protoconid and extends as a strip along the buccal crest of the metaconid as far as the tip of that cusp. It faces posteriorly, upwards, and somewhat buccally.

Facet 6 occupies the edge of the posterior crest of the paracone of the upper molar (precentrocrista). On the lower molar, the facet is on the buccal slope of the cristid obliqua. The upper facet faces posteriorly, downwards, and somewhat lingually; and the lower facet faces anteriorly, upwards, and somewhat buccally.

Facets 2 and 6 are produced when the paracone of the upper molar travels down the groove between the protoconid and the oblique crest of the lower molar [the hypoflexid of VAN VALEN, 1966] (fig. 3). At the same time, the paraconule pas-

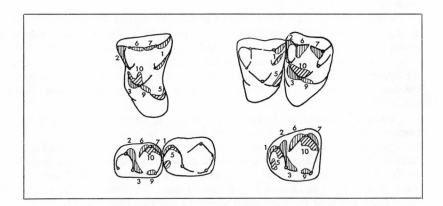

Fig. 4. Left: *Palenochtha minor,* USNM 9590, 9647. M^1, M_{1-2}. Right: *Pronothodectes simpsoni,* U. Wyo. 1099, Carn. Mus. 1699. M^{1-2}, M_2. Wear facets shaded to show direction of striations.

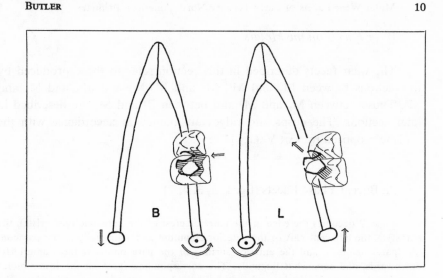

Fig. 5. Diagram to illustrate the production of wear striations in two directions, as postulated by MILLS. In B (buccal phase), the mandible is rotating around the ipsilateral condyle, with the result that the lower molar crosses the upper molars transversely, producing the striations shown. In L (lingual phase), the mandible is rotating round the contralateral condyle, resulting in oblique striations.

ses from near the tip of the metaconid along its buccal crest (fig. 7). In centric occlusion (fig. 6), the paracone tip is near the buccal margin of the lower molar; and the paraconule is near the notch in the protoconid-metaconid crest (thus near the anterior end of the oblique crest).

Facet 7 forms on the anterior crest of the metacone of the upper molar (postcentrocrista), and on the lower molar it occupies the posterior face of the hypoconid. It is parallel to facet 2. Facets 6 and 7 are produced as the hypoconid of the lower molar passes through the notch between the paracone and the metacone of the upper molar. In centric occlusion (fig. 6), the tip of the hypoconid is opposite the deepest part of the trigon basin, approximately between the paraconule and the metaconule.

Facet 1 is formed on the posterior crest of the metacone (metacrista) and the posterior crest of the metaconule (when present). It is parallel to facet 6; i.e., it faces posteriorly, downwards, and somewhat lingually. The corresponding lower facet is situated on the anterior crest of the protoconid and may extend lingually to involve the anterior surface of the paraconid when that cusp is sufficiently developed. Wear on the anterobuccal cingulum of the lower molar is also part of this facet. The lower facet faces anteriorly, upwards, and somewhat buccally.

Facets 7 and 1 are produced by the movement of the metacone down the groove between the talonid of one lower tooth and the trigonid of the more posterior tooth. The tip of the metacone first makes contact with the hypoconulid, and in

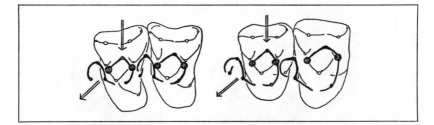

Fig. 6. M^{1-2}/M_{1-2} in centric occlusion. Left: *Palenochtha minor,* USNM 9590, 9647. Right: *Plesiadapis jepseni,* USNM 20781, 20586. Contact points indicated by shaded circles. Arrows indicate directions of movement of the hypoconid of M_1 in the buccal and lingual phases.

centric occlusion (fig. 6) it lies in the embrasure between two lower teeth. During the same movement, the metaconule travels from a position anterior to the paraconid to touch the hypoconulid in centric occlusion.

Facets 1 and 2 are produced by movement of the protoconid from a position near the parastyle lingually along the paracingulum of the upper molar to reach its centric position anterior to the paraconule.

Facet 3 occupies the anterior surface of the protocone, extending along its anterior ridge (preprotocrista) and involving also the lingual part of the anterior cingulum (precingulum). It faces anteriorly, downwards, and somewhat lingually and is thus parallel to facet 1. On the lower molar, facet 3 occupies the posterior surface of the metaconid, including the posterior ridge of that cusp from which a metastylid may develop; and it extends along the posterior face of the trigonid below facet 2. Facet 3 faces posteriorly, upwards, and slightly buccally, like facet 2; but the two facets are usually not in the same plane, being separated by a ridge or narrow shelf on the posterior face of the trigonid (fig. 7). This shelf represents the path of the tip of the paraconule; it is sometimes formed by a lingual extension of the cristid obliqua towards the metaconid, as in plesiadapids and *Pelycodus.*

Facet 9 is a small area on the posterior surface of the protocone near its tip. It occupies part of the ridge that in most early primates develops on the posterior or posterolingual side of the protocone. It faces posteriorly, downwards, and lingually. The corresponding facet on the lower molar occupies the anterobuccal surface of the rather low entoconid; it faces anteriorly, upwards, and buccally.

Facets 3 and 9 are formed by the passage of the protocone through the notch between the metaconid and the entoconid, buccally and downwards into the talonid basin. At the same time, the metaconid moves from a position near the paraconule lingually along the precingulum.

The true hypocone is too small in the early primates to make any contact with the entoconid (facet 8); but in some specimens of *Notharctus* a small facet is present on the anterolingual surface of the pseudohypocone, and there is a corresponding facet on the posterobuccal face of the entoconid (fig. 8). The pseudohypocone, which arises from the posterior ridge of the protocone, is not homologous with the

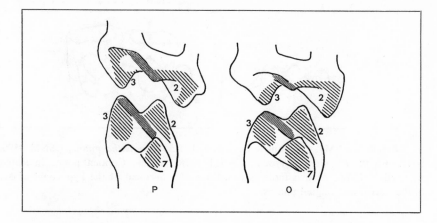

Fig. 7. P=*Palaechthon alticuspis*; O=*Omomys carteri*. Anterior surface of an upper molar and posterior surface of a lower molar, orientated so that striations are parallel. Path of protoconule shown in heavier shading. Note that in *Palaechthon* the movement has a greater vertical component than in *Omomys*.

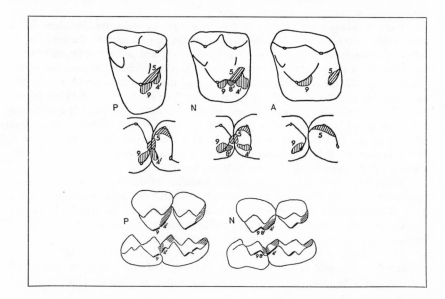

Fig. 8. Diagrams to illustrate occlusal relations of the pseudohypocone of Notharctidae. P=*Pelycodus frugivorus*; N=*Notharctus tyrannus*; A=*Adapis parisiensis* (for comparison). The lowest figures show M^{2-3}/M_{2-3}, orientated so that the buccal-phase facets are seen in profile; facet 5 (and its equivalent on the hypoconulid of M_3) shaded.

true hypocone, which arises from the cingulum. Its entoconid contact may be called facet 8'. It does not occur in *Adapis*.

A more important function of the pseudohypocone, however, seems to be its contact with the paraconid, as was noticed by GREGORY [1920, fig. 61]. In *Pelycodus* a small facet occurs on the posterior surface of the protocone, lingually to the metaconule and posterobuccally to facet 9 (fig. 8). It forms on an indistinct posterobuccal ridge of the protocone and faces more posteriorly than facet 9. It apparently occludes with the anterior surface of the paraconid, having taken over the function of the metaconule in this respect. It may be called facet 4'. In *Notharctus* this facet occupies the posterior surface of the pseudohypocone. In the more advanced forms of *Notharctus* the paraconid retreats towards the metaconid, especially on M_3; and facet 4' may be transferred to the metaconid, thus imitating the hypocone-metaconid relation (facet 4) found in higher primates.

B. The Buccal-Phase Movement

Striations on the facets so far described indicate that all are produced by a single upward and medial (lingual) movement of the lower jaw. This is confirmed by the fact that by rotation of the teeth it is possible to find a direction of view from which all the facets can be seen tangentially at the same time (fig. 3). The movement is evidently the same as the 'power stroke' observed in *Didelphis* by CROMPTON and HIIEMÄE [1970]. There is a considerable transverse component. As seen in crown view, the direction of movement is nearly perpendicular to the line of the teeth, as if its center lay near the ipsilateral jaw joint as in living primates (fig. 5). When seen in anterior view, scratches show that the direction of movement is approximately that of a line drawn through the tips of the paracone and the paraconule, or of the metacone and the metaconule (fig. 7). When the tooth is seen along the direction of movement, the conules disappear behind the paracone and metacone. In the lower molars the direction of movement is approximately that of a line drawn through the tip of the metaconid and the bottom of the notch near the anterior end of the oblique crest.

In *Didelphis* the direction of movement is at about 35 ° to the vertical in the earlier stages of occlusal contact, falling to about 45 ° as the lower teeth pass more lingually [CROMPTON and HIIEMÄE, 1970, fig. 7C]. It is not possible to give accurate comparable figures for fossil primates, owing to uncertainty about the orientation of teeth in relation to the vertical. If it is assumed that the roots of the lower molars are vertical, the angle may be estimated at about 45 ° in most Middle Palaeocene primates

(Palaechton, Palenochtha, Pronothodectes), comparable with that in *Didelphis*; but it increases in later forms with less elevated cusps to about 55 °, as in *Pelycodus, Notharctus, Plesiadapis, Omomys, Tetonius,* and *Phenacolemur* (fig. 7). This interpretation contrasts with that of SZALAY [1968a], who believes that a marked transverse shear in molar occlusion is a primitive character, reduced in primates. He bases his view on the presence in primitive eutherians, such as Palaeoryctidae, of long transverse crests, accompanied by a comparatively lingual position of the paracone and metacone in the upper molar. However, the direction of wear scratches in Palaeoryctidae is quite steep, forming an angle of about 40 ° to the vertical, and comparable with *Didelphis*. The relationship between the angle of shear and the position of the cusps is complex and in some cases may be the opposite of what Szalay supposes. As MILLS [1954] has pointed out, tilting of the shearing planes of the buccal cusps into a less vertical position would carry the tips of the cusps buccally in the upper molar and lingually in the lower molar, thus reducing the length of the paracrista and similar crests.

C. Lingual-Phase Facets (fig. 2, 3, 4)

When all the foregoing buccal-phase facets are made to disappear from view by suitable orientation of the tooth, some other facets still remain. Striations on these additional facets are not parallel to those of the buccal-phase facets, but indicate a different direction of jaw movement, distinguished by MILLS [1955] as the lingual phase of occlusion.

Facet 10 forms on the buccal surface of the protocone, extending along its anterior ridge (preprotocrista). It also involves a rounded ridge that in many primates runs directly buccally from the tip of the protocone towards the bottom of the trigon basin. The facet faces downwards and slightly buccally. The corresponding facet of the lower molar occupies the lingual surface of the hypoconid, extending along the hypoconid-hypoconulid crest and also anteriorly along the edge of the oblique crest. In many forms the hypoconid possesses a short lingual ridge that assists in the formation of the facet. The orientation is upwards and slightly lingual. Striations show that during the formation of facet 10 the tip of the hypoconid moves anterolingually, from its centric position in the center of the trigon basin toward a notch in the preprotocrista immediately lingual to the paraconule (fig. 6). The notch is worn by the oblique crest of the lower molar that travels through it. At the same time, the protocone travels posterobuccally across the talonid basin toward a notch between the hypoconid and the hypoconulid worn by the oblique crest of the upper molar. During this movement, the lingual surface of the hypoconid and the buccal surface of the protocone slide across each other; and it is of interest to note that these surfaces are very frequently roughened by minor enamel-ridges, which tend to be arranged across the direction of movement. Such ridges can therefore function as does the roughened surface of a millstone.

Facet 5 occupies the lingual part of the posterior cingulum of the upper molar. In forms in which the cingulum is continuous with a posterior crest of the proto-cone, the facet continues onto this crest, where it is situated basally to the entoconid contact (facet 9). When a hypocone differentiates, facet 5 occupies its buccal sur-face; and this situation is true also of the pseudohypocone of *Notharctus*. The facet faces downwards and somewhat anteriorly and buccally. The corresponding facet on the lower molar occupies the lingual surface of the protoconid, near its tip, and ex-tends for a short distance along the protoconid-metaconid crest. What appears to be a detached portion of facet 5 occurs at the tip of the paraconid in some forms *(Pronothodectes, Plesiadapis, Phenacolemur, Tetonius)*. In *Pelycodus* and *Notharc-tus* the facet involves the anterior margin of the trigonid rather than the paraconid itself, and it extends forward to the hypoconulid of the more anterior tooth (fig. 8). On the lower molar, facet 5 faces upwards and somewhat anteriorly and lingually. In the production of the facet the trigonid moves forwards and lingually, in a direc-tion approximately parallel to the oblique (protocone-metaconule) crest of the upper molar. The tip of the protoconid passes along the posterolingual cingulum of the upper molar, and the development of the hypocone from this cingulum enables the contact to be maintained when the trigonid of the lower molar is reduced in height. This seems to be the only function of the hypocone of early primates, unlike the pseudohypocone of Notharctidae.

D. The Lingual-Phase Movement

As seen in crown view, the lingual phase has a considerable propalin-al component (fig. 6). The direction of movement is approximately per-pendicular to the protocone-paraconule crest (preprotocrista) of the up-per molar and parallel to the oblique crests of the upper and lower mo-lars. It thus makes an angle of 45 ° or more with the direction of buccal-phase movement. This angle increases from the first to the third molars, in a manner consistent with the location of a center of lingual-phase rota-tion near the condyle of the opposite side of the head. There is also a con-siderable difference in a vertical direction between the two movements. Buccal-phase movement includes an important jaw-closing (orthal) com-ponent, so that at its end the teeth are maximally interlocked in the posi-tion of centric occlusion. Lingual-phase movement, starting from centric occlusion, involves a small degree of opening of the jaws. But the move-ment is mainly a horizontal one; in a vertical plane through the direction of motion, the inclination is roughly estimated at 10 ° to 15 ° below the horizontal. Thus, the lingual phase is in no way a continuation of the buccal phase, but an entirely different method of moving the jaw (fig. 5).

E. The Centric Position (fig. 6)

The two movements intersect in the centric position. The relations between upper and lower molars are then as follows: the paracone and metacone stand laterally to the buccal margin of the lower tooth, the paracone laterally to the hypoflexid, and the metacone laterally to the interdental embrasure; the protocone stands posteriorly to the metaconid and opposes the deepest part of the talonid basin; the protoconid is in the interdental embrasure, anterior to the paraconule; the metaconid is near the precingulum, anterior to the protocone; the hypoconid opposes the deepest part of the trigon basin; the entoconid stands posteriorly and lingually to the tip of the protocone. None of these relations results in a close contact between the teeth such as would act as a stop to bring the power stroke to an end. This was noticed by Crompton and Hiiemäe [1970, p. 32]: '... the tips of most of the cusps do not come into contact with the opposing teeth during shearing occlusion.' In primates it appears that the contacts that stop the power stroke are those between the crests of the hypoconid and the crests of the protocone. There are two of these contacts, an anterior one, where the oblique crest of the lower molar meets the preprotocrista near the protoconule, and a posterior one, where the posterior hypoconid crest near the hypoconulid meets the oblique crest of the upper molar near the metaconule. Corresponding contacts appear to exist in *Didelphis*, to judge from figure 5A$_2$ of Crompton and Hiiemäe [1970], even though the conules are rudimentary or absent.

Third Molars (fig. 9)

Characteristic of early primates is the 'third lobe' of M_3, usually considered as an enlarged hypoconulid. It functions as a substitute for the missing trigonid of M_4, and its relations to M^3 are similar to those of the trigonid of M_3 to M^2. When the hypoconulid of M_3 is sufficiently enlarged, a facet on its anterior surface occludes in the buccal phase with a facet on the posterior surface of the metacone of M^3, like facet 1 (e.g., *Phenacolemur, Plesiadapis, Pelycodus, Notharctus*). On the lingual sur-

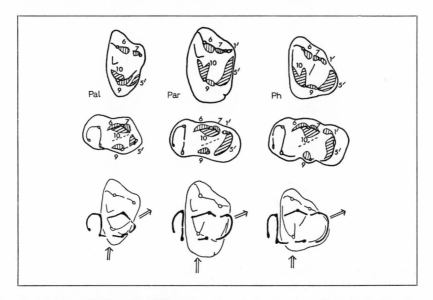

Fig. 9. Occlusal relations of the talonid of M_3. Pal = *Palenochtha minor*, USNM 9590, 9639 (entoconid restored); Par = *Paromomys maturus*, USNM 9540, 9473; Ph = *Phenacolemur praecox*, Princ. U. 13028 (reconstruction based on SIMPSON, 1955, pl. 33), USNM 19168. Lowest figures show the position of the cusps in centric occlusion. Arrows indicate directions of relative movement of the protocone during the buccal and lingual phases.

face of the hypoconulid is a facet produced in the lingual phase, as shown by its oblique striations. It represents a contact with the posterior cingulum of M^3 and may be compared with facet 5, the height of the hypoconulid compensating for the lack of a hypocone on M^3. Especially in more heavily worn teeth, a groove may be seen crossing the talonid basin of M_3 in a posterobuccal direction and leading towards the notch between the hypoconid and the hypoconulid. This represents the path of the protocone of M^3 during the lingual phase, and the buccal slope of the groove forms facet 10.

Premolars (fig. 10)

The premolars of early primates show a greater variety of structure than the molars, and this is reflected in their occlusal relations. Although

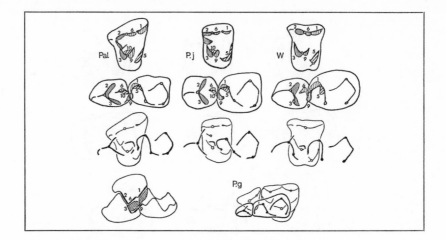

Fig. 10. Occlusal relations of P⁴. Pal = *Palenochtha minor*, AMNH 35443 (rev.), 35451; P.j = *Plesiadapis jepseni*, USNM 20781, 20586 (rev.); W = *Washakius insignis*, USNM 17795, U. Wyo. 1319 (rev.). Top row, crown views of left P⁴. Second row, crown views of right P_4 and M_1, with facets that occlude with P⁴ shaded. Third row, P⁴, P_4, and M_1 in centric occlusion. Bottom row, left, *Palenochtha minor*, P⁴ P_4, and M_1 seen in direction of buccal-phase movement. P.g=*Plesiadapis gidleyi*, AMNH 17372, Dm³ and Dm⁴ (rev.) in centric occlusion with Dm_3 and Dm_4, to show relation of the paraconid of Dm_4 to the protocone of Dm³.

their wear facets are simplified, they may in general be compared with those of the molars. As premolars depart from the molar pattern, the number of wear facets is reduced; the last to disappear are facets 1 and 2.

The posterior part of P⁴ has relations with the trigonid of M_1. In the buccal phase the anterior and buccal side of the M_1 trigonid occludes with the lingual and posterior side of the P⁴ metacone or, when this cusp is absent, with the corresponding part of the posterior crest of the P⁴ paracone (facet 1). The metaconule of P⁴ is usually absent.

P⁴ protocone often contacts the anterior surface of M_1 paraconid during the buccal phase. This facet seems to be an extension of, or replacement for, facet 9, as the entoconid of P_4 is reduced and overhung by the forwardly-projecting paraconid of M_1. It is probably because of this function that the paraconid of M_1 frequently remains in a forward position, while that of M_2 retreats towards the metaconid.

On the reduced talonid of P_4 it is possible to find facet 6 (P⁴ paracone against P_4 oblique crest), but facet 7 (P⁴ metacone against P_4 hypoconid) is usually absent owing to the weak development of the cusps concerned. Facet 3 (P^4 protocone against P_4 metaconid) is always present, even when the metaconid is represented only by the posterolingual crest of the protoconid. Similarly, facet 2 (P_4 protoconid against the anterior surface of P^4) is always present. In Plesiadapidae its area is increased by the paraconule-like cusp of P^4.

In the lingual phase, the protoconid of M_1 occludes with the posterior cingulum of P^4; and sometimes (e.g., *Washakius*) a small hypocone is differentiated on the cingulum, similar in function to that of M^1 (facet 5). Contact between the P^4 protocone and the P_4 hypoconid (facet 10) is frequently present in a reduced form, but it is lost in those genera in which the talonid of P_4 is particularly reduced.

Carpolestidae (fig. 11)

The Carpolestidae are a Palaeocene group characterized by the specialization of their premolars, which have been compared with those of multituberculates. The upper premolars differ so markedly from the lowers that it is not immediately obvious how the teeth fit together. Carpolestids are generally regarded as primates; SIMPSON [1940] noted the resemblance of their lower molars to those of Plesiadapidae. It should, therefore, be possible to relate carpolestid occlusal relations to those of other early primates.

The following specimens were examined:

Elphidotarsius florencae Gidley: USNM 9411 (the type, showing the lower dentition);
Carpolestes dubius Jepsen: Princeton University 14077 and 19349 (upper dentitions; I am indebted to Prof. G. I. JEPSEN for supplying casts of these specimens);
Carpolestes sp.: Carnegie Museum 11518 (M_1), 11557 (M_1 and M_2), 11524 (P_4);
Carpodaptes hazelae Simpson: AMNH 33930 (associated upper and lower dentitions);
Carpodaptes aulacodon Matthew and Granger: AMNH 17367 (the type, with P_4–M_2).

Molar occlusal relations appear to resemble those of *Pronothodectes*. The buccal-phase movement is fairly steep, making an angle of about 45 ° to what is estimated as the vertical. Oblique wear scratches are indistinct, but were seen on the buccal surface of the protocone of M^2 of *Carpodaptes hazelae*; and the shape of the talonid of M_3 is so similar to that of other primates that it was almost certainly used in the oblique move-

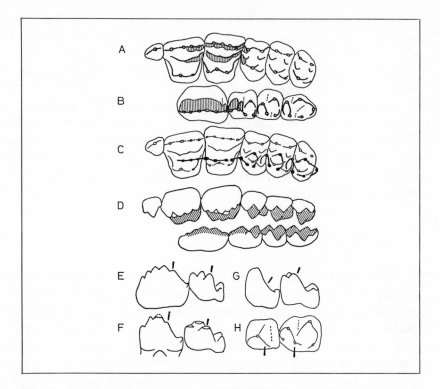

Fig. 11. A–E *Carpodaptes hazelae,* AMNH 33980 (lower teeth reversed). *A, B* Crown views P³–M³ and P₄–M₃; buccal-phase facets of the shearing ridge of P₄–M₁ shaded. *C* Diagram to show centric occlusion. *D* The dentitions seen in direction of relative movement; lingual-phase contact area shaded. *E* Lingual view of P₄ and M₁. *F Elphidotarsius florencae,* lingual view of P₄ and M₁. *G, H Pronothodectes matthewi,* USNM 9847, P₄ and M₁ in lingual and crown views. In E–H, the metaconids are indicated.

ment. Lingual-phase function seems to have been best developed on the third molars and progressively less so on the more anterior teeth.

In *Carpolestes* and *Carpodaptes,* M₁ differs from that of other early primates in that the arms of the trigonid are more widely divergent, the metaconid being posteriorly displaced and the paraconid situated directly anterior to the protoconid. *Elphidotarsius* shows a lower degree of specialization in the same direction. A large facet 1 on the buccal surface of the trigonid occluded with P⁴.

P⁴ and P³ (unknown in *Elphidotarsius*) have three rows of cusps. The buccal row consists of five subequal cusps (the anterior one, probably a parastyle, reduced on P³) that form a high, serrated, longitudinal ridge. Lingually to these is a less prominent intermediate ridge, in line with the conules of the molars and poorly differentiated into cusps. The lingual margins of P⁴ and P³ are occupied by a more basal ridge on which stand three cusps (the central one is apparently the protocone); the anterior cusp is poorly developed on P³. The edges of the buccal and intermediate ridges are worn to form buccal-phase facets that face lingually and downwards.

P₄ consists essentially of a high longitudinal ridge, the edge of which is differentiated into cusps. *Elphidotarsius* shows that this ridge has been formed by opening out the trigonid, thus bringing the metaconid almost directly posterior to the protoconid. In *Elphidotarsius*, an accessory cusp has developed on the anterior arm of the protoconid; and the number of accessory cusps increases in *Carpodaptes* and *Carpolestes* to produce a serrated cutting edge. This edge continues the line of the trigonid of M₁.

As the jaws closed in the buccal phase, the ridge formed by P₄ and the anterior part of M₁ passed up the lingual surface of P³ and P⁴, shearing against the edges of the buccal and intermediate ridges of the upper teeth. The wear surfaces so formed would correspond to facets 1 and 2. M₁ would occlude with the posterior part of P⁴, and P₄ would occlude with the anterior part of P⁴ and most of P³. The movement would be brought to an end by the shelf formed by the lingual row of cusps on the upper premolars. Owing to the narrowness of this shelf, no significant lingual-phase function occurs.

The intermediate ridge of the upper premolars may be regarded as an extension of the protoconule-like cusp that develops on the premolars of some Plesiadapidae, where it takes part in the formation of facet 2. However, in Plesiadapidae the metaconid of P₄ is very rudimentary or absent, being replaced functionally by a posterolingual ridge of the protoconid, which contacts the protocone of P⁴ to form facet 3. In Carpolestidae, the metaconid of P₄ is high, taking part in the main crest; and it has no contact with the protocone, except possibly when the crest of P₄ meets the lingual ledge of P⁴ at the end of the shearing stroke. Although a common ancestor of the two families is possible, it is unlikely that the ancestral morphology of P₄ was like that of Plesiadapidae. *Saxonella*, from the Palaeocene of Germany [RUSSELL, 1964], does have such a P₄; and for this reason it would seem to fall better into the Plesiadapidae than into the Carpolestidae. Its referred P⁴ is not greatly different from that of *Pronothodectes*, and its upper incisors are also plesiadapid in character.

The Significance of the Lingual Phase

The majority of wear facets on the teeth of early primates correspond to those found by CROMPTON and HIIEMÄE [1970] in *Didelphis* and were presumably produced by the movement described by those authors as the 'power stroke'. The main difference noted is that in many primates the movement is less vertical and more transverse than in *Didelphis*.

There remain, however, some wear facets that cannot be accounted for by this type of movement. These include particularly wear on the buccal surfaces of the protocone and the hypocone and on the lingual surfaces of the protoconid and the hypoconid. The movement that produces these is directed obliquely and inclined only slightly from the horizontal. It can take place only when the lower molars are positioned to the lingual side of the centric occlusal position. It involves a forward (or backward) sliding of the ipsilateral condyle and appears to have a center of rotation on the opposite side of the head, near the contralateral condyle. This condition is the lingual phase of occlusion, as defined by MILLS [1955] (fig. 5).

Two functional explanations of the lingual phase are given by MILLS. On the assumption that centric occlusion coincided with a symmetrical (centric) relation of the lower jaw to the skull, the lingual phase on one side of the mouth would take place at the same time as the buccal phase on the other. On the active side, the lower molars would travel lingually and upwards in relation to the upper molars to reach the position of centric occlusion. MILLS [1963] describes this as a 'cusp-in-groove action' which 'would serve to cut or slice the food'. On the balancing side, the lower molars would at the same time move buccally and posteriorly to reach the centric position; this lingual-phase relation is not regarded as an active chewing function, but one which 'may help to reduce the stress on the mandibular joint'. On the other hand, if the lingual phase occurs on the active side, it would involve a movement forward and lingual from the centric position, i.e., in the reverse direction to that involved in balancing occlusion, 'producing a grinding action analogous to a millstone'. The buccal phase would then be carried out in reverse on the balancing side, the lower molars moving buccally from the centric position.

In *Didelphis*, CROMPTON and HIIEMÄE [1970] showed that when the molars are working on one side of the mouth, those on the other side are quite out of contact. Centric occlusion occurs when the mandible is displaced laterally from its symmetrical position. Moreover, the two rami of the mandible move on each other at the symphysis; and the condyles approach each other during the power stroke. There is, therefore, no occlusal balance in *Didelphis*. The open symphysial suture of many primitive primates suggests that some inter-ramal mobility may have been present; moreover, in *Tarsius* and Tupaiidae, MILLS [1963, 1967] found evidence of a transverse condylar movement (Bennett shift) that might imply an approach of the condyles toward each other during chewing. In Cercopithecidae and in *Indri*, which have specialized bilophodont molars, MILLS [1955, 1963] found that the lingual phase is reduced, contact between the teeth taking place mainly in the buccal phase; during much of the buccal phase on the working side, the teeth are out of contact on the balancing side. Although this condition is undoubtedly a derivative one, it suggests that occlusal balance may not be a function of great importance.

As an alternative to providing occlusal balance, the lingual phase might be considered as a distinct mode of chewing. CROMPTON and HIIEMÄE object to this interpretation because they are unable to see how pressure between the teeth could be maintained. They assume that during the lingual phase the tip of the hypoconid would pass over the tip of the protocone, as seems to be implied by some of MILLS' diagrams [e.g., 1955, fig. 13]. This path would necessitate some degree of opening of the jaws. However, if the hypoconid passes through a notch between the protocone and the paraconule, the movement is more nearly horizontal. It could be produced by the external pterygoid muscle, assisted by the internal pterygoid and external masseter, while pressure between the teeth was maintained by the vertical fibers of the masseter and temporal muscles. The small degree of jaw opening still necessary could be brought about by the reaction of the teeth themselves. The function would, as MILLS suggested, be essentially one of grinding, including the breaking up of seeds and insects. It would not be an integral component of the chewing cycle; for after centric occlusion is reached at the end of the buccal phase, the mouth could be opened and the buccal phase repeated. The lingual phase would be a second form of chewing that could be called in as occasion required.

The Lingual Phase in Other Mammals (fig. 12)

Is the lingual phase peculiar to primates? To answer this question a survey was made of Early Tertiary mammals, seeking for wear facets on

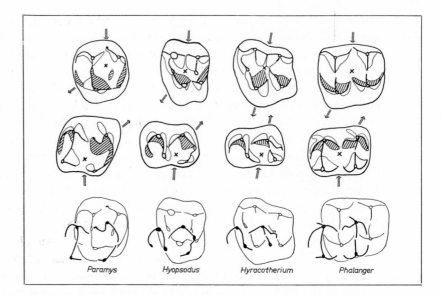

Fig. 12. Molar wear facets of four mammals with evidence of a lingual phase of occlusion. Upper left and lower right molars, and diagrams to show centric occlusion. × indicates position of the hypoconid and protocone in relation to the upper and lower molars, respectively, in centric occlusion. The arrows show the relative movement of these cusps in the buccal and lingual phases of occlusion, as evidenced by striations.

the buccal surfaces of the protocone and the hypocone and on the lingual surfaces of the hypoconid and the protoconid (and the hypoconulid of M_3) that could not be produced by the power stroke movement. The criteria for such facets were that they should be plane surfaces with striations. No such facets were found in Apatemyidae, Picrodontidae [confirming SZALAY, 1968b], Leptictidae, Pantolestidae, Plagiomenidae, or lipotyphlous insectivores [contrary to the finding of MILLS, 1966]. They also appear to be absent in Palaeoryctidae, Oxyaenidae, Hyaenodontidae, Miacidae (and all Fissipeda), and Mesonychidae, as well as in Periptychidae and some Arctocyonidae such as *Arctocyon*. In these forms, wear of the protocone and hypoconid is essentially abrasive; and there is no evidence of shearing or grinding action between these cusps, although wear facets produced by the power stroke are easily seen. On the other hand, lingual-phase facets were seen in Mixodectidae *(Eudaemonema)*, Rodentia, Condylarthra *(Hyopsodus, Tricentes, Tetraclaenodon, Meniscotherium)*, Perissodactyla, Litopterna, *Coryphodon,* and some Artiodactyla (Dichobunidae). They also occur in some recent Phalangeridae.

This distribution strongly suggests that lingual-phase occlusion was not a primitive characteristic of tribosphenic dentitions but has been evolved as an adaptation to a herbivorous or omnivorous diet. In recent Tupaiidae, though scratches show that some lingual-phase movement occurs, it is of small extent; for only a small facet 10 develops on the lingual part of the preprotocrista. In older specimens, the crests of the protocone are worn by abrasion, and the facet is no longer distinguishable. Facet 5 apparently does not develop, as the lingual movement is insufficient to bring the protoconid into contact with the lingually placed hypocone.

In the more primitive rodents (e. g., *Paramys, Ischyromys,* Aplodontidae), there are two sets of facets, one produced by a transverse movement corresponding to the buccal phase and the other by an obliquely anterior movement centered apparently around the contralateral condyle, as in primates. The facets of these primitive rodents may, therefore, easily be compared with those of primitive primates, a fact of some interest in view of a suggested relationship between the two orders. In the more advanced rodents, the lingual-phase facets become of greater importance at the expense of the buccal-phase facets, resulting in the characteristic propalinal chewing action. In rodents there is no doubt that the lingual phase is an active method of grinding and not merely a form of occlusal balance.

In ungulates, the angular change of direction of movement between buccal and lingual phases, as seen in crown view, is generally smaller than in primates or rodents. From about 30° in *Hyopsodus* and *Tricentes,* the angular change in direction is reduced in Perissodactyla and Artiodactyla almost to zero, so that the lingual phase appears to be a continuation of the buccal phase, except that the lower teeth are rising in the buccal phase and falling in the lingual phase. In these advanced ungulates, the hypoconid does not pass, as in primates, through the protocone-paraconule notch, but passes instead more directly across the protocone. In *Hyracotherium,* for example, the tip of the hypoconid passes slightly behind the tip of the protocone; and the scratches on facet 10 are transverse instead of being perpendicular to the protoloph. A similar situation occurs in Dichobunidae; but in the higher artiodactyls the protocone is displaced forwards or (in Caenotheriidae) backwards, and the lingual phase is largely eliminated.

This variability of the lingual phase contrasts with a remarkable uniformity of buccal-phase relations and supports the view that the lingual phase was an addition to the occlusal repertoire of tribosphenic teeth. It may have evolved at least twice in placentals, once in the primate-rodent stock and again in the condylarth-ungulate stock, as well as in phalangerid marsupials. The associated adaptations of the condyles and jaw musculature remain to be worked out.

The lingual phase has nothing to do with the 'grinding mode' of occlusion postulated by VAN VALEN [1966, p. 47] in Palaeoryctidae, in which the trigonid is said to move across the trigon in various ways. VAN VALEN does not describe the facets in detail and apparently does not distinguish between true wear facets and wear of the tips of cusps due to food abrasion. My own observations of *Cimolestes, Puercolestes, Gelastops, Acmeodon,* and *Didelphodus* are that the more upstanding parts of the teeth in this group were particularly subject to food abrasion and were probably used to hold relatively large food objects; all the striated wear facets are explicable by a single buccal-phase power stroke like that of *Didelphis,* corresponding to VAN VALEN's 'shearing mode'.

References

BUTLER, P. M.: The milk-molars of Perissodactyla, with remarks on molar occlusion. Proc. zool. Soc., Lond. *121:* 777–817 (1952a).

BUTLER, P. M.: Molarization of the premolars in the Perissodactyla. Proc. zool. Soc., Lond. *121:* 819–843 (1952b).

BUTLER, P. M.: Relationships between upper and lower molar patterns. Int. Colloq. on the evolution of lower and non-specialized mammals. Kon. Vl., Acad. Wet. Lett. Sch. Kunst, België; part I, pp. 117–126 (Brussels 1961).

BUTLER, P. M. and MILLS, J. R. E.: A contribution to the odontology of *Oreopithecus.* Bull. br. Mus. natur. Hist., Geol. *4:* 1–26 (1959).

CROMPTON, A. W. and HIIEMÄE, K.: How mammalian molar teeth work. Discovery, Lond. *5:* 23–24 (1969).

CROMPTON, A. W. and HIIEMÄE, K.: Molar occlusion and mandibular movements during occlusion in the American opossum *Didelphis marsupialis* L. J. Linn. Soc. (Zool.) *49:* 21–47 (1970).

CROMPTON, A. W. and JENKINS, F. A.: Molar occlusion in late Triassic mammals. Biol. Rev. *43:* 427–458 (1968).

DAHLBERG, A. A. et KINZEY, W.: Etude microscopique de l'abrasion et de l'attrition sur la surface des dents. Bull. Grpmt. int. Rech. scient. Stomat. *5:* 242–251 (1962).

EVERY, R. G.: The teeth as weapons. Their influence on behaviour. Lancet *i:* 685–688 (1965).

EVERY, R. G. und KÜHNE, W. G.: Funktion und Form der Säugerzähne. I. Thegosis, Usur und Druckusur. Z. Säugetierk. *35:* 247–252 (1970).

GREGORY, W. K.: The origin and evolution of the human dentition. Part II. J. dent. Res. *2:* 215–283 (1920).

MACINTYRE, G. T.: The Miacidae (Mammalia, Carnivora). I. The systematics of *Ictidopappus* and *Protictis*. Bull. amer. Mus. natur. Hist. *131:* 115–210 (1966).

MILLS, J. R. E.: The dental occlusion of the primates; M.Sc. thesis Manchester (1954).

MILLS, J. R. E.: Ideal dental occlusion in the primates. Dent. Pract. *6:* 47–61 (1955).

MILLS, J. R. E.: Occlusion and malocclusion of the teeth of primates; in BROTHWELL Dental anthropology, pp. 29–51 (Pergamon Press, London 1963).

MILLS, J. R. E.: The dentitions of *Peramus* and *Amphitherium*. Proc. Linn. Soc., Lond. *175:* 117–133 (1964).

MILLS, J. R. E.: The functional occlusion of the teeth of Insectivora. J. Linn. Soc. (Zool.) *47:* 1–25 (1966).

MILLS, J. R. E.: A comparison of lateral jaw movements in some mammals from wear facets on the teeth. Arch. oral Biol. *12:* 645–661 (1967).

RUSSELL, D. E.: Les mammifères paléocènes d'Europe. Mem. Mus. nat. Hist. natur. N. S., C *13:* 1–324 (1964).

SIMPSON, G. G.: Studies on the earliest primates. Bull. amer. Mus. natur. Hist. *77:* 185–212 (1940).

SIMPSON, G. G.: The Phenacolemuridae, new family of early primates. Bull. amer. Mus. natur. Hist. *105:* 411–442 (1955).

STONES, H. H.: Oral and dental diseases; lst ed. (Livingstone, Edinburgh 1948).

SZALAY, F. S.: The beginnings of primates. Evolution *22:* 19–36 (1968a).

SZALAY, F. S.: The Picrodontidae, a family of early primates. Amer. Mus. Novit. *2329:* 1–55 (1968b).

VAN VALEN, L.: Deltatheridia, a new order of mammals. Bull. amer. Mus. natur. Hist. *132:* 1–126 (1966).

ZINGESER, M. R.: Sexual dimorphism in monkey canine teeth. Proc. 8th Int. Congr. of Anthropological and Ethnological Sciences, vol. 1, pp. 305–308 (Council of Sciences of Japan, Tokyo 1968).

ZINGESER, M. R.: Cercopithecoid canine tooth honing mechanisms. Amer. J. phys. Anthrop. *31:* 205–213 (1969).

ZINGESER, M. R.: The prevalence of canine tooth honing in primates. Annu. Meet. Amer. Ass. phys. Anthropol., Boston 1970; cit. Amer. J. phys. Anthrop. *35:* 300 (1971).

Author's address: Dr. P. M. BUTLER, Department of Zoology, Royal Holloway College, University of London, *Englefield Green, Surrey* (England)

Symp. IVth Int. Congr. Primat., vol. 3: Craniofacial Biology of Primates,
pp. 28–64 (Karger, Basel 1973)

Evolutionary Trends in the Dynamics of Primate Mastication[1]

KAREN M. HIIEMÄE and R. F. KAY

Unit of Anatomy with special relation to Dentistry, Guy's Hospital Medical
School, London, and Museum of Comparative Zoology, Harvard University;
and Divison of Vertebrate Paleontology, Peabody Museum, Yale University,
and Museum of Comparative Zoology, Harvard University

Introduction

Answers to the question 'how are teeth used' and its corollary 'how is
their shape and position in the jaw related to this use' do not loom large
in the voluminous literature of dental anthropology. General statements
abound, but the few specific studies undertaken have generally deduced
or inferred the use (as distinct from feeding behaviour) from tooth morphology and wear patterns [BUTLER, 1952; BUTLER and MILLS, 1959;
MILLS, 1955, 1963, 1967]. The hazards inherent in this approach have
been discussed elsewhere [HIIEMÄE, 1967; CROMPTON and HIIEMÄE,
1970]. It should be emphasized that teeth and dentitions have evolved in
response to changing functional demands, and not the reverse. Morphological trends, whilst shedding light on primate phylogeny [see, e.g.,
FRISCH, 1965] themselves reflect changes in the pattern of use of the dentition in the living animal.

This paper is concerned with the use of the dentition in four species
of extant genera, *Tupaia, Galago, Saimiri,* and *Ateles,* and will draw general conclusions on the functional and evolutionary trends underlying the
obvious differences in the morphology of the teeth and jaws. Two methods have been used. Normal jaw function in feeding has been recorded ci-

1   This research was supported by the United States Public Health Service, National Institutes of Health grant DE-02648 (K. H.), and the Wenner-Gren Foundation for Anthropological Research grant 2735 (R. F. K.).

nefluorographically and the recordings analyzed to give information on the pattern of jaw movement and the feeding behaviour of each animal. The detailed results of these studies will be presented elsewhere [KAY and HIIEMÄE, 1973a], only the basic pattern and timing of the jaw movement being reported here. The shape of the molar teeth in each species has been examined and related to its function during the power stroke of mastication; fossil materials have also been examined and related to this study. The conclusions to be drawn from both studies are then assessed. During the course of these investigations some general principles have emerged, as has the necessity for a clear understanding of what is meant by a number of the terms found indispensable by authors in this field. It is therefore essential to preface a description of our results by a short review of what is meant by 'use' and to examine its structural implications when applied to the dentition.

The mammalian dentition evolved by the gradual addition of a new activity, 'food preparation' or *mastication*, to the primitive 'food handling' or *ingestive* function found in most reptilian and other vertebrate dentitions. With the addition of this new stage to the feeding process between ingestion and deglutition, the mammalian molar dentition evolved; and specialisation of these teeth occurred. Within the structural framework demanded for the three basic stages of feeding, accessory functions were assumed, the most important in primates being those involved in social interaction and communication. Man must be the only mammal to spend more 'functional time' using the jaw apparatus for purposes other than that for which it originally evolved.

Only when brought into contact with food or each other do the teeth come into action and then only passively, as preparators or triturators of food. The lower dentition approaches, contacts, and traverses the upper as the result of jaw movements produced by the integrated activity of the jaw muscles. These muscles also control the position of the jaw with some assistance from the temporo-mandibular joint. When the teeth actually come into contact, their shape controls the direction of lower jaw movement for the duration of that contact, the 'contact' or 'glide' movement of the envelope of motion. In animals so far studied, this period occupies a surprisingly small percentage of the total time spent in feeding [CROMPTON and HIIEMÄE, 1970; HIIEMÄE and CROMPTON, 1971; KAY and HIIEMÄE, 1973a].

Although ingestion and mastication are separate stages in the feeding process, there is some overlap in the patterns of movement. In ingestion the food is being transferred into the oral cavity as a bite of suitable size for triturition. (A *bite* is the unit of food separated in ingestion and reduced in mastication.) This may or may not involve some separation of the bite from its matrix, i.e., 'incision' or 'ingestion by mastication'. Whereas only some types of ingestive behaviour such as suckling, lapping, licking, and ingestion by mastication involve rhythmic and cyclical movements, all mastication is based on the repetition of a three-stroke movement cycle [MØLLER, 1967; HIIEMÄE and ARDRAN, 1968; CROMPTON and HIIEMÄE, 1970; HIIEMÄE and KAY, 1973; KAY and HIIEMÄE, 1973a]. The masticatory cycle is defined as having three strokes, preparatory, power, and recovery, and begins at the point of maximum gape. [These terms are fully defined in KAY and HIIEMÄE, 1973a; but also see HIIEMÄE and CROMPTON, 1971.] Some authors discussing mastication in man [e.g., BEYRON, 1964] use a different terminology and different criteria, referring to 'opening' and 'closing' phases without distinguishing a separate power stroke. Further, the reference point used to distinguish the beginning or end of a cycle is either the 'rest position' or that of 'centric occlusion' or 'centric relation'. For technical reasons, not least the difficulty of determining these positions cinefluorographically in experimental animals, these points have not been used.

The preparatory stroke brings the lower teeth into position for the power stroke in which the food is triturated; at its completion, the lower jaw is then carried downwards and away during the recovery stroke, passing from occlusion to maximum gape. As far as can be ascertained [HIIEMÄE and CROMPTON, 1971], the number, amplitude, duration, and type of each cycle or sequence of cycles is related to the consistency of the food undergoing trituration. There are two basic types of masticatory cycle, characterized by the type of power stroke involved: 'puncture-crushing', in which the food is pulped and the teeth do not come into contact; and 'chewing', where the food is reduced by a variety of methods and where tooth contact probably occurs. It is this second type of power stroke that depends on contact movements guided by the shape of the teeth; and it, in turn, governs the effect of the teeth on the food held between them.

Most accounts of mastication, and this will be no exception, are almost exclusively concerned with the occlusal relations of the teeth, the movements of the lower jaw, and in some cases, the activity or biomechanics of the muscles. It is not the purpose of this paper to present a detailed account of feeding, including dietary habits and food handling behaviour, as well as of masticatory, movement, in the primates studied. There is now the beginnings of a literature based on field studies of the subject [HLÀDIK and HLÀDIK, 1969; CHARLES-DOMINIQUE, 1971], and this has been augmented by some experimental studies [e.g., KAY and HIIEMÄE, 1973a]. However, it is necessary to emphasize that, although this report is primarily concerned with the trends to be observed in the pattern of jaw movement and the occlusal relationships of the teeth in a series of four extant mammals, the process by which food is introduced into the mouth, is reduced, and then swallowed is a function of much more than the skeletal elements and their associated musculature. The hands of primates collect, prepare, and transfer food to the mouth. Further, tough food is often gripped between the cheek teeth, the head pulling in one direction, the hand in another. These two opposing forces, coupled with the compressive force exerted on the food through the teeth, separate the bite from its matrix. Once in the mouth, although the bite is reduced by the interaction of upper and lower teeth, it is positioned between them by the combined actions of the lips, cheeks, and tongue. Moreover, the relative movement of the teeth is produced not only by movement of the lower jaw but also by cranial extension and flexion involving the cervical spine and the pre- and post-vertebral musculature. In effect, any description of the feeding process restricted, as in this case, to only two of the stages in an elaborate series of events [see HIIEMÄE, 1967] explores only one aspect of a complex system. This deficiency is all the more marked when the data obtained are used to extrapolate and, by analogy, to explain the behaviour of fossil forms.

Materials and Methods

A. Cinefluorography

The technique and the apparatus used have been described elsewhere by CROMPTON [1968], CROMPTON and HIIEMÄE [1970], and KAY and HIIEMÄE [1973a]. Specimens of *Tupaia glis* and *Ateles* sp. were obtained from commercial suppliers. We are greatly indebted to Dr. B. TRUM and Dr. F. GARCIA of the New England Regional Primate Research Center for making available the specimens of *Galago crassicaudatus* and *Saimiri sciureus* used in this study and for aiding in the insertion of radio-opaque markers. Such markers (see fig. 1), which greatly facilitate analysis of the cinefluorographic recordings, were inserted into the jaws of *Galago, Saimiri,* and *Ateles* under Nembutal anaesthesia. This insertion was not possible for the specimens of *Tupaia*, which proved extremely difficult to handle. After suitable acclimatisation to the experimental conditions, the animals were fed a variety of foods; and cinerecordings[2] were made at a camera speed of 60 frames per second (with Eastman Kodak Plus X negative film) in the projections and at the kilo-voltages and milliamperages shown in table I. The cinerecordings were then ana-lysed by the methods described in CROMPTON and HIIEMÄE [1970] and by HIIEMÄE and CROMPTON [1971] and as indicated in the text.

B. Occlusal Analysis

Specimens of *T. glis, G. crassicaudatus, S. sciureus,* and *Ateles* sp. from the osteological collection at the Museum of Comparative Zoology (MCZ), Harvard Uni-versity, have been examined to determine the morphology of the teeth, the extent and definition of the wear facets, and the geometrics of occlusal relations during the power stroke. In addition, the following have been used for comparative purposes: the upper and lower first molars of *Palenochtha minor* (AMNH[3] 35443 and AMNH 35450) from the Middle Paleocene, Gidley Quarry, Montana; the upper and lower molars of *Pelycodus trigonodus* (YPM[4] 25012 and YPM 23317) from the Early Eocene, Big Horn Basin, Wyoming; upper and lower first molars of *Aegypto-pithecus zeuxis* (YPM 23798 and YPM 21032) from the Middle or Late Eocene, Quarry I, Fayum Province, Egypt.

The nomenclature used in the descriptions of the teeth examined is that devel-oped by VAN VALEN [1966] and SZALAY [1969] and later modified by CROMPTON [1971]. Where appropriate, terms from the primate literature have also been used.

Observations

Two separate but interrelated investigations have been undertaken, one into feeding behaviour and masticatory movements in species of *Tu-paia, Galago, Saimiri,* and *Ateles* and a second into the molar form and occlusion of these animals and an equivalent fossil series *(Palenochtha, Pelycodus,* and *Aegyptopitheus).* The results obtained are described un-der two general headings: 'Cinefluorographic Studies' and 'Molar Form and Occlusion'. The following report is not an exhaustive statement of our results, but rather an account in which emphasis is placed on such trends as can be discerned in the evolution of primate molars. Detailed

---

2   Sections from these recordings have been combined to form a short cinefilm, 'Mastication in Primates'.
3   AMNH = American Museum of Natural History.
4   YPM = Peabody Museum, Yale University.

Fig. 1. Lateral view radiographs (natural size) showing the position of the radio-opaque markers used to facilitate analysis of the cinefluorographic recordings. A=*Ateles* sp.; B=*Saimiri sciureus*; C=*Galago crassicaudatus*.

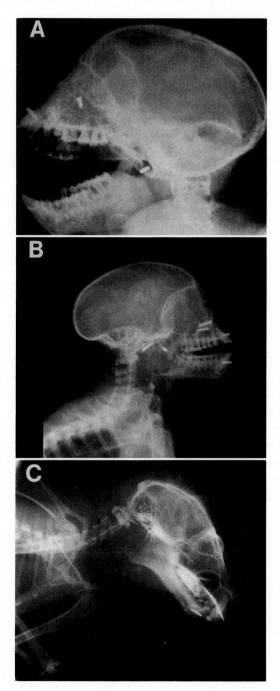

Table I. Food supplied and projections used in the cinefluorographic recordings. The kilovoltage (first figure) and milliamperage (second figure) used in each projection are given

	Food	Projection		
		lateral	dorso-ventral	frontal
Tupaia	salami, apple, grape		in the range 58–70/15–20	70/16
Galago	apple	58/16	58/22	70/16
	biscuit biscuit and water	60/16	60/16	70/1[6]
Saimiri	apple biscuit biscuit and water	60/16	65/16	84/16
Ateles	apple biscuit	80/18	110/20	100/20

accounts of the masticatory behaviour and occlusion of each extant species are either in press or in preparation.

Mention has already been made of the terminological problem associated with descriptions of feeding behaviour. Two pairs of expressions will frequently be used in the following account: 'tooth–tooth' and 'tooth–food–tooth' contact; and 'chewing' and 'puncture–crushing'. The first pair refers to the probable or discernible relationship of the cheek teeth, the second to the manner in which the food is undergoing trituration. When a bite of food is first masticated, its bulk separates the occlusal surfaces of upper and lower cheek teeth; such contact is, therefore, 'tooth–food–tooth' contact. The jaw movements associated with the initial trituration of such a bite are characterized as 'puncture–crushing' since they involve the penetration of the tips of the cusps into the material (puncturing) as it is compressed (crushed) between the opposing molar surfaces. Similarly, once the food has been sufficiently broken down for 'chewing' to take place, then at some stage in the power stroke 'tooth–tooth' contact will, or can, occur. It must be emphasised that these terms are generalisations insofar as chewing may well involve only tooth–food–tooth contact, but they serve to distinguish two distinct types of feeding behaviour and their consequential wear on the teeth.

Table II. The use of the hands and of the various sections of the tooth row in ingestion and mastication by species of *Tupaia, Galago, Saimiri,* and *Ateles*

	Ingestion		Mastication		
	hands	incisors	premolars	puncture-crushing (premolars-molars)	chewing (molars)
Tupaia	yes, one usually	no, holding/grasping	yes ingestion by	yes mastication	regular, few side changes
Galago	yes, one usually	no, holding/grasping	yes ingestion by	yes mastication	regular, side changes every 6 or more cycles
Saimiri	yes, one, or for large objects, both	yes, soft food	yes hard food only	yes large numbers of cycles on active side	regular, side changes about every 3 cycles
Ateles	yes, one, or for large objects, both	yes, soft food	yes hard food only	yes large numbers of cycles on active side	regular, side changes about every 3 or more cycles

A. Feeding Behaviour

All four species (which for convenience will hereafter be referred to only by their respective genera) were hand-fed and studied for considerable periods; table II shows in summary form the general behavioural observations made. The incisors are not actively used in biting by either *Tupaia* or by *Galago*, but are used for grasping and holding. All major activity that requires the generation of masticatory forces such as biting or chewing is a function of the cheek teeth in these animals. Ingestion of this type uses movements identical with those of puncture–crushing and has been called 'ingestion by mastication' [HIIEMÄE and CROMPTON, 1971]. *Saimiri* and *Ateles* use their incisors for biting, although very hard food is transferred to the cheek teeth for ingestion.

This shift from premolar-molar incision to incisor incision is obviously correlated with the development of spatulate incisors in the higher primates and can be regarded as a specialisation of the anterior tooth row, since the restricted use of the incisors in feeding by the prosimians studied [KAY and HIIEMÄE, 1973a] is also found in the primitive mammal *Didelphis* [HIIEMÄE and CROMPTON, 1971]. It is therefore suggested that the absence of a positive ingestive function for the incisors in these animals and, by analogy from the available data on *Tupaia* and *Didelphis*, in primitive mammals provided the opportunity for adaptations in the anterior teeth not necessarily associated with feeding and resulted in the wide variety of incisal adaptations found in extant mammals. Incisors may, therefore, have non-nutritive functions, as in the dental combs of lemurs and lorises, or may exhibit food-collection adaptations, as do those of *Daubentonia* or rodents [HIIEMÄE and ARDRAN, 1968].

It seems possible that the mandibular symphysis, demonstrably mobile in species of extant forms like *Tupaia* (see below), *Galago* [KAY and HIIEMÄE, 1973a], and *Didelphis* [COMPTON and HIIEMÄE, 1970], fused phylogenetically at approximately the same time as the spatulate incisor evolved. Spatulate incisors and a fused symphysis, coupled with canines projecting beyond the occlusal plane, are characteristic of both platyrrhines and catarrhines. The extent to which the two features are linked remains a subject for further investigation; but, since a new and considerable load was placed on the incisors for the first time in primate history, it seems unlikely that the two events were unconnected.

B. The Masticatory Cycle

A masticatory sequence in a mammal involves repeated cycles, each cycle consisting of three strokes. The cycle is regarded as beginning at the point of widest opening (maximum gape) and consists of an upstroke (the preparatory stroke), followed by the actual triturition or power stroke, and completed by a downstroke (recovery stroke). A puncture-crushing cycle is distinguished from a chewing cycle by a slowing of the upstroke as the teeth begin to make contact with the food and by a limited degree of mandibular movement in the power stroke. The recovery stroke of one cycle may merge smoothly into the preparatory stroke of the next, or there may be an abrupt transition. A summary of the pattern of movement in a typical cycle for *Tupaia, Galago, Saimiri,* and *Ateles* species follows. In all four animals, cranial flexion and extension on the vertebral column forms an essential part of the movement; such relative elevation of the upper jaw is most marked where the mouth is being widely opened,

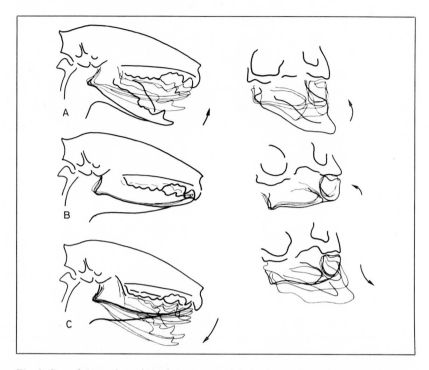

Fig. 2. Superimposed tracings from sequential single frames from cinerecordings of *Tupaia glis* taken in the lateral (left) and antero-posterior (right) projections. The head is slightly skewed in the antero-posterior projection so that the body of the mandible on the animal's left (the active side) is seen in coronal section. A=preparatory stroke; B = power stroke; C = recovery stroke.

as in ingestion or puncture-crushing, but occurs in all but the smallest gapes.

Despite the basic differences in jaw architecture (*Tupaia* and *Galago* species have mobile symphyses, relatively long V-shaped tooth rows, and virtually flat occlusal planes; species of *Saimiri* and *Ateles* have fused symphyses, more U-shaped dental arcades, and slightly curved occlusal planes), the pattern of movement of the lower jaw in the masticatory cycle is basically the same in all four species studied (fig. 2, 3, 4).

1. The Preparatory Stroke

The first stroke of the cycle begins with the lower jaw positioned to the active side of the midline (fig. 4) and with the mandible maximally de-

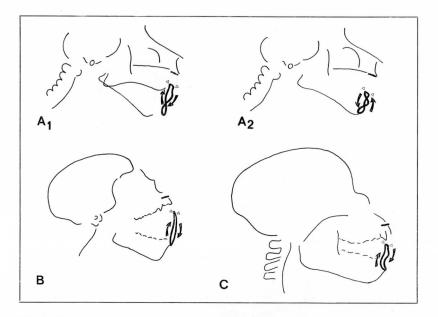

Fig. 3. The path of movement of the lower jaw marker in sequential single frames from lateral projection cineradiographs of *Galago crassicaudatus* in a puncture-crushing cycle (A_1) and a chewing cycle (A_2), showing the increase in the antero-posterior component and diminution of the vertical movement in the latter; of *Saimiri sciureus* (B) and *Ateles* sp. (C) in a chewing cycle. The path of movement shown is obtained by joining the consecutive marker positions from the series of frames; the solid arrows show the direction of movement, the hollow triangles the cinefluorographic limits of the power stroke.

pressed for that cycle. Maximum gape is not a constant position. It reflects the degree of mandibular depression at that time, although in each animal the gape reached tends to be much the same for successive cycles when the same type of trituration, i.e. puncture-crushing or chewing, is in progress (fig. 5, 6). From this lateral depressed position, the lower jaw is then elevated until tooth–food–tooth contact is reached on the active side. The degree of medio-lateral excursion involved in this stroke varies and is most marked in *Tupaia* (fig. 4), where, near the bottom of the stroke, there is some degree of 'overswing' that is corrected by a medial movement as it reaches completion.

In *Galago, Saimiri*, and *Ateles* species, the upward movement of the preparatory stroke is complicated by some antero-posterior movement (fig. 3). As the jaw is elevated, it is first retracted and then protruded as

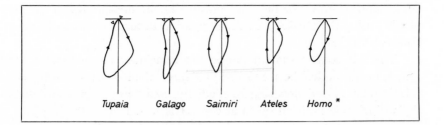

Tupaia   Galago   Saimiri   Ateles   Homo *

Fig. 4. A diagrammatic representation of the path of movement of a point on the lower jaw in the antero-posterior projection through a typical chewing cycle where the active side is on the animal's right (left in figure). The horizontal line represents maximal occlusion; the vertical, a perpendicular to the occlusal plane at centric occlusion. The hollow arrows indicate the beginning and end of the power stroke in the experimental animals.

* The profile for *Homo sapiens,* taken from AHLGREN [1966], represents the simplest of a large number of possible variations, most of which can be correlated with malocclusions.

the stroke comes to completion. This point is of some importance, since GINGERICH [1971] has suggested that the terminal stages of the preparatory stroke and the beginning of the power stroke (i.e., the stage in the cycle where puncture-crushing occurs) may involve a definite retraction and, in fact, did so in *Adapis.* This subject will be discussed further below. The extent to which the essentially vertical movement of the preparatory stroke is complicated by antero-posterior and medio-lateral components appears to be correlated with masticatory behaviour. Puncture-crushing cycles are, in general, much simpler than chewing cycles (fig. 3) and have a greater vertical dimension. In all four animals, the stroke is completed with alignment of the lateral (buccal) surfaces of the upper and lower molars on the active side.

2. The Power Stroke

In all four species studied, the power stroke begins with the buccal surfaces of the molars on the active side in vertical alignment. The lower jaw is then moved upwards, medially, and slightly forward to bring the teeth on the active side into centric occlusion [*sensu* CROMPTON and HIIEMÄE, 1970]. This is phase I. In *Tupaia,* the last part of this first stage involves a slightly more marked anterior component than at the beginning and also a definite rotation of the lower jaw about its long axis. (This feature can be seen in the cinefluorographic recordings by the change in the dorso-ventral profile of the jaw and also in the appearance of the symphysis during the stroke.) A similar rotation occurs in *Galago* [KAY and HIIEMÄE, 1973a].

There is some variation during the remainder of the stroke. Once centric occlusion has been reached in *Tupaia*, the lower jaw on the active side is then moved downwards, medially, and toward the contralateral (balancing) side until all occlusal contact is lost. During the last part of this movement, transitory balancing-side contact may occur, but without any alteration in the smoothness or direction of movement. In *Galago*, the lower jaw on the active side moves downwards and medially from centric occlusion and then further toward the balancing side so that the molars on the balancing side come into transitory contact. This excursion is immediately followed by a return movement toward the midline and downward into the recovery stroke [KAY and HIIEMÄE, 1973a]. This peculiar additional component interspersed between the power stroke (*sensu strictu*) and the recovery stroke has not been observed in any other animal. There is no detectable balancing-side contact in *Saimiri*; however, such contact can be seen in *Ateles*. When the molars on the active side reach centric occlusion, the lingual cusps on the balancing side come into alignment. Viewed in antero-posterior projection radiographs, this contact is clearly cusp to cusp; and the whole mandible appears slightly tilted in the coronal plane. As the molars on the active side move out of occlusion, a downward, forward, and medial movement, this balancing-side contact ceases.

The power stroke in *Tupaia*, *Galago*, *Saimiri*, and *Ateles* thus involves two movements: the initial (and probably the only functional phase in puncture-crushing) upward, medial, and forward movement to centric occlusion (phase I); and a second downward, forward, and medial movement from centric occlusion until occlusal contact is lost (phase II). During phase I, there is no balancing-side contact in any of the animals studied. In phase II, as far as can be seen from the cinefluorographic recordings, non-functional transitory balancing-side contact occurs in *Tupaia* and in *Ateles*. There is a specific movement toward the balancing side to achieve this contact in *Galago* and no contact in *Saimiri*.

3. The Recovery Stroke

The recovery stroke begins (by definition) when occlusal or tooth–food–tooth contact is lost on the active side and is completed when the lower jaw is depressed to maximum gape. During this predominantly vertical movement, lateral and antero-posterior excursions occur (fig. 2, 3) so that the mandibular symphysis moves toward the balancing side prior to returning to the active side before, or as, maximum gape is reached. If a side change is about to occur, the direction of the recovery stroke is correspondingly altered. Instead of returning back across the midline to maximum gape on the active side, the lower jaw continues laterally and downwards on the balancing side (now the active side); and a new preparatory stroke on that side follows.

In all the animals studied, the downward movement of the recovery stroke is often halted at a vertical position of about 5° of gape (fig. 5, 6) or, as in *Ateles,* after a clear freeway space has developed. During this 'hold', although vertical movement is suspended, a simple protrusive movement occurs. At its completion, rapid depression (fig. 5) follows until maximum gape is reached. If no 'hold' occurs, mandibular movement in the recovery stroke is initially very slow until reaching the 'hold'-gape dimension and then continues rapidly.

The foregoing general description of masticatory behaviour in species of *Tupaia, Galago, Saimiri,* and *Ateles* illustrates the essential similarity between the movements involved in all four species and, as can be seen from figure 4, between them and man. There is some variation in the amplitude and direction of movement within successive cycles in each animal; but in general, as can be seen from figures 5 and 6, behaviour in successive cycles of the same type (i.e. puncture-crushing or chewing) is remarkably consistent. Figure 4 is diagrammatic insofar as it shows a single 'typical' cycle in each case. However, the profile for *Homo sapiens,* taken from AHLGREN [1966], represents the simplest of a large number of possible variations, most of which can be correlated with malocclusions.

A masticatory sequence begins with the initial puncture-crushing of the bite. The number of puncture-crushing cycles involved in the 'softening' process varies with the consistency of the food; but all except the softest food is 'pulped', if only for one or two cycles, before the chewing cycles commence. As a general rule, the gapes are larger; and the power strokes involve smaller medio-lateral and antero-posterior excursions in puncture-crushing cycles. Once the food has been reduced to a consistency suitable for chewing, then a longer sequence of chewing cycles begins and continues until deglutition. *Ateles* showed an interesting variation; instead of the normal sequence of P-C (puncture-crushing) followed by Ch (chewing) cycles, the sequence is quite often P-C, P-C, Ch, Ch, side change P-C, P-C, Ch, Ch, Ch. The regularity and rhythmicity of the mas-

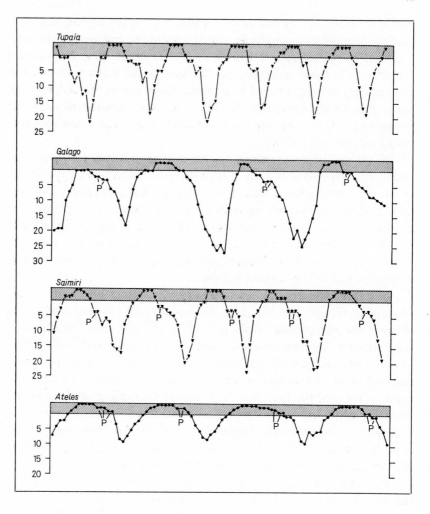

Fig. 5. Plots of gape (expressed in degrees on the vertical axis) and time (horizontal axis) for a series of masticatory cycles in *Tupaia glis*, *Galago crassicaudatus*, *Saimiri sciureus*, and *Ateles* sp. The shaded area corresponds to the range of mandibular depression between complete closure (top) and clear freeway space (bottom). Each point of the plots corresponds to the position of the lower jaw in a single frame; thus, the horizontal axis is equivalent to about 1.5 sec elapsed time. The letter 'p' on the two lower plots indicates that simple protrusion is occurring. The slope of the plot indicates the rate of movement, which, with the exception of *Ateles* sp., is maximal in the later stages of the recovery stroke.

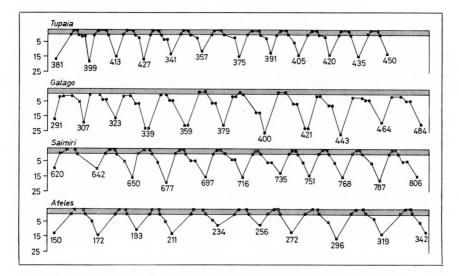

Fig. 6. Gape/time plot for the experimental animals used to illustrate the regularity of jaw movement in successive cycles. In this figure, the horizontal axis corresponds to about 3 sec elapsed time, and only the gapes at the beginning of each stroke and at the 'hold' have been measured (in degrees). The numbers inserted at each maximum are the numbers of cinefluorographic recording frames (taken at 60 f.p.s.) counted between the respective maxima.

ticatory sequence is shown in figures 5 and 6, where time (expressed as the number of frames of 60 f.p.s. cinefluorographic recordings) is shown on the horizontal axis and gape (expressed in degrees from a wide freeway space) is shown on the vertical axis. As the regularity was not restricted to the rate and direction of movement, but also applied to the duration of successive cycles, as illustrated by the frame numbers noted on figure 6, a statistical examination of the timing of the masticatory cycle has been undertaken.

C. The Duration of the Masticatory Cycle

Previous studies [HIIEMÄE and CROMPTON, 1971] have shown that the absolute duration of the masticatory cycle as well as the percentage duration of its constituent strokes can vary with the types of food and with the manner in which they are undergoing trituration. This variation can be examined by using each frame of the cinerecording (taken at 60 f.p.s.) as a measure of time.

With the use of the 'single shot' mode on the analysing projector, frame counts have been made for a large number of cycles (n varies from 93 to 111) from lateral projection recordings of each of the four animals. The beginning and end of each stroke were determined by 'overshooting' at least one frame; and the following criteria were used to define each stroke.

The *preparatory stroke* begins at the first upward movement from maximum gape and is completed at either tooth–tooth contact (readily visible in the cinerecordings) or at the cessation of regular upward movement on contact with the food. This point is less easily determined and is of some importance in puncture-crushing cycles, where the preparatory and power strokes form a continuum with an intermediate hiatus in vertical movement. In such a stroke, the maximum elevation of the jaw may not be reached until the end of the power stroke.

The *power stroke* commences at tooth–tooth or tooth–food–tooth contact and is completed when a freeway space can first be clearly identified. This represents a marginal extension of the stroke but is readily discernible in the recordings.

The *recovery stroke* commences with the appearance of a clear freeway space and is completed at the point of maximum gape.

All the counts were made by one observer: the counting error was one frame in 20. Each sequence of cycles was counted from the completion of ingestion to either deglutition or to the next 'space' on the film. (The apparatus used limits filming to 20-sec bursts.) Where unequivocal puncture-crushing occurred, this was noted for the cycle. The frame counts were then treated as follows: the total time for each cycle was expressed as the total number of frames counted for that cycle, and the duration of each of the constituent strokes (in frames) was expressed as a percentage of the total cycle time. The mean and standard deviation for the total and percentage-stroke times for (a) all cycles, (b) puncture-crushing, and (c) chewing cycles were calculated and, where relevant, the significances of the differences between the mean values found. The results for the four species are shown in table III. From these figures table IV was prepared to allow interspecific comparisons. The results are discussed in relation to the tables.

1. Tupaia (table IIIA)

As can be seen in table III, the mean overall duration of the masticatory cycle is 14.65 ± 2.1 frames; but when it is broken down into puncture-crushing and chewing cycles, their means are found to be significantly different ($p < 0.05$). This difference in the length of the cycles was not observed in any of the other animals. *Tupaia* also differs from *Galago, Saimiri,* and *Ateles* in that the preparatory stroke is of the same percentage duration in both puncture-crushing and chewing cycles; whilst the means of the power and recovery strokes are very highly significantly different ($p<0.001$). In calculating the significance levels of the mean percentage durations of the strokes, no weighing was given to the difference in the absolute durations of the two types of cycle. Even so, the relative significances involved ($p<0.001$ as opposed to $p<0.05$) are such that the significance of the difference in stroke duration greatly outweighs that of cycle time. The essential differences between the two types of cycle can be shown by expressing the mean percentage duration of each of their constituent strokes as a ratio. If the preparatory stroke is taken as unity, since for this species it is of the same relative duration for both types of cycle, the respective ratios are approximately:

Puncture-crushing 1:1:0.9
Chewing 1:0.8:1.4.

However, in all the other animals examined it is the power stroke that is of the same relative duration in both types of cycle; for comparative purposes, therefore, the ratios are also given taking the power stroke as unity:

Puncture-crushing 1:1:1.1
Chewing 1.3:1:1.8.

2. *Galago* (table IIIB)

There is no significant difference between the mean durations of the cycle types in *Galago*, nor in the percentage duration of the types of power stroke. Very highly significant differences ($p<0.001$) are, however, found in the percentage durations of the preparatory and recovery strokes. The former is much longer in puncture-crushing than in chewing cycles, and the latter is proportionally reduced. This situation is reflected in the ratios of their percentage durations, taking the power stroke as unity, namely:

Puncture-crushing 1.6:1:1.2
Chewing 1.2:1:1.9.

3. *Saimiri* (table IIIC)

The results for *Saimiri* show the same pattern as for *Galago*, but the squirrel monkey is unique among the four species examined in having a preparatory stroke consistently shorter than either its power or recovery

Table III. Mean total cycle times and mean percentage durations of each stroke in all masticatory cycles together and in puncture-crushing and chewing cycles separately for *Tupaia, Galago, Saimiri,* and *Ateles*. Where significant differences occur, the pairs of figures involved are bracketed together and the probability figures given; all other data are expressed as means ± standard deviations

	Total cycle time[1]	Total cycle time, %		
		preparatory stroke	power stroke	recovery stroke
A. Tupaia				
n = 111	14.65 ± 2.19	31.08 ± 5.6	27.40 ± 7.0	41.33 ± 7.3
Puncture-crushing n = 48	15.16 ± 2.3 } p<0.05	31.85 ± 4.9 } p<0.001	31.35 ± 5.9 } p<0.001	36.74 ± 6.4
Chewing n = 63	14.30 ± 1.8	31.13 ± 4.6	24.45 ± 5.9	44.83 ± 6.0
B. Galago				
n = 94	18.53 ± 3.5	30.93 ± 8.5	23.77 ± 6.5	44.76 ± 6.3
Puncture-crushing n = 27	18.37 ± 3.3 } p<0.001	37.68 ± 6.0 } p<0.001	22.84 ± 7.2 } p<0.001	40.32 ± 6.8
Chewing n = 66	18.87 ± 2.8	29.15 ± 1.5	24.16 ± 6.2	45.56 ± 5.1

Table III (continued)

C. Saimiri				
n = 104	21.42±3.4	28.72±5.6	33.18±2.2	38.07±5.4
Puncture-crushing				
n = 37	20.27±4.8	31.57±6.2 ⎱ p<0.001	33.22±6.3	35.07±5.7 ⎱ p<0.04
Chewing				
n = 67	21.45±3.8	26.15±8.7	33.22±5.0	38.30±1.3
D. Ateles				
n = 100	19.66±3.03	36.13±7.2	28.36±6.1	34.88±7.9
Puncture-crushing				
n = 34	19.35±4.2	43.70±5.1 ⎱ p<0.001	29.39±8.2	26.78±4.3 ⎱ p<0.001
Chewing				
n = 66	19.53±3.2	32.23±1.4	28.30±5.2	39.06±5.9

1 Expressed as numbers of cinerecording frames at 60 f.p.s.

strokes. Despite this proportional difference, the preparatory stroke is longer in puncture-crushing cycles and the recovery stroke shorter than in chewing cycles; the power stroke is approximately the same in both types. The relevant ratios are:

Puncture-crushing 0.9:1:1
Chewing 0.8:1:1.1.

4. *Ateles* (table IIID)

The pattern seen in *Galago* and *Saimiri* species is also found in *Ateles;* a clear distinction (p<0.001) occurs in the percentage duration of the preparatory and recovery strokes, but no such differences in the absolute cycle time or in the percentage duration of the power stroke are seen. The ratios are:

Puncture-crushing 1.6:1:0.9
Chewing 1.1:1:1.4.

5. *Analysis of Results*

These results reflect the behaviour of one representative of each species over a large number of masticatory cycles (n ranges from 94 to 111). In all cases, puncture-crushing and chewing cycles are clearly distinguished not, save for *Tupaia*, by differences in the absolute duration of each cycle, but by the relative proportions of the cycle time used for each stroke. The preparatory stroke is relatiely lengthened in *Galago, Saimiri,* and *Ateles* and the recovery stroke shortened. *Tupaia* is distinctive in that the power stroke and not the preparatory stroke is lengthened. Observation suggests that, although this is a real difference, its biological significance, let alone its taxonomic value, is probably small. *Tupaia* has sharply pointed, high-cusped cheek teeth; by using the criteria outlined above, which are cinefluorographically justifiable, to define the two strokes, frames may be counted in the power stroke that are actually showing the terminal stages of the preparatory stroke. The two strokes form a continuum, the change marked by a reduction in the rate of upward movement and a shift to medial and anterior movement. This 'turn-point' is clear on antero-posterior radiographs for low-cusped teeth, but is much less easy to define in *Tupaia*.

As far as possible, variation in the masticatory behaviour of each animal has been considered by using large numbers of cycles. Previous studies [HIIEMÄE and CROMPTON, 1971], where numbers of animals were

Table IV. The interspecific variation in the absolute duration of the masticatory cycles and in the percentage duration of each constituent stroke. The data are expressed as the means±standard deviations of the sample sizes given

	Total cycle time[1]	Total cycle time, %		
		preparatory stroke	power stroke	recovery stroke
A. All cycles				
Tupaia (n=111)	14.65±2.09	31.08±5.6	27.40±7.0	41.33±7.3
Galago (n=94)	18.53±3.3	30.93±8.5	23.77±6.5	44.76±6.3
Saimiri (n=104)	21.42±3.4	28.72±5.6	33.18±2.2	38.07±5.4
Ateles (n=100)	19.66±3.03	36.13±7.2	28.36±6.1	34.88±7.9
B. Puncture-crushing cycles				
Tupaia (n=48)	15.16±2.3	31.85±4.9	31.35±5.9	36.74±6.4
Galago (n=27)	18.37±3.3	37.68±6.0	22.84±7.2	40.32±6.8
Saimiri (n=37)	20.27±4.8	31.57±6.2	33.22±6.3	35.07±5.7
Ateles (n=34)	19.35±4.2	43.70±5.1	29.39±8.2	26.78±4.3
C. Chewing cycles				
Tupaia (n=63)	14.30±1.8	31.13±4.6	24.45±5.9	44.83±6.0
Galago (n=67)	18.87±2.8	29.15±5.3	24.16±6.2	45.56±5.1
Saimiri (n=67)	21.45±3.8	26.15±8.7	33.22±5.0	38.30±1.3
Ateles (n=66)	19.53±3.2	32.23±1.4	28.30±5.2	39.06±5.9

1 Expressed as frames of cinerecordings taken at 60 f.p.s.

used, showed little intraspecific variation in feeding behaviour. Such variation is, however, well documented in man, where a single individual may show marked fluctuation in the form and direction of movement through a cycle; and even greater variation is found between individuals. As many of the factors alleged to affect the chewing pattern in man, such as social awareness, convention, malocclusion, etc., do not apply or only minimally apply to the experimental animals used, it is felt that interspecific comparisons using the data in table IV are justified.

Table IV presents the results for each species aggregated in three sections: all cycles (A), puncture-crushing cycles (B), and chewing cycles (C). Each section has been separately examined by the technique of analysis of variance in order to explore the following possibilities: that each species is distinctive (species effect); that the proportions of the strokes within the cycles are distinctive (cycle effect); and that there is a significant interaction between species and cycle effects (residual effect).

In all three sections, the species effect is negligible, i.e. not significant. This finding means that the variation seen is not likely to be greater than could have occurred had the results been obtained from four animals of the same species rather than four of different species. When the variation in the percentage duration of the strokes in each section of the table is considered, the cycle effect is found to be significant in section B and very significant in section C ($p < 0.01$). In short, the distribution of the percentage stroke times over the two types of cycle (B and C) stands out as a distinctive entity independent of the species. The residual effect, which is almost entirely the interaction between cycle and species, reflects the fact that each form has its own characteristic variance in percentage stroke times and that these are not constant over the four types. For all three sections, all cycles, the cycle/species interaction is very significant (section A, $p < 0.01$; sections B and C, $p < 0.001$). The only conclusion for section B, puncture-crushing cycles, is that the residual effect, the interaction or combination of species and cycle characteristics, is the only important feature, as is also the case for chewing cycles (section C), where the cycle effect itself stands out as a very significant feature.

These results show that although the mean values given in the tables vary, that variation lies within the cycle for each species, rather than between species. *Tupaia, Galago, Saimiri*, and *Ateles* all use their own variants of a standard pattern for puncture-crushing and chewing. It can be concluded that changes in the morphology of the masticatory apparatus in general, and of the cheek teeth in particular, have not involved any significant change in the pattern of mastication as expressed by absolute cycle times or the percentage durations of each of the strokes; the time base has remained more or less constant. Alterations in both the form of the jaw apparatus and, therefore, of the mode of chewing may well have occurred within a pre-existing framework.

D. Occlusion in the Power Stroke

Cinefluorographic studies of feeding allow the patterns of movement and the time base of mastication to be investigated, but only under optimal conditions can the relationships of the cheek teeth during the power stroke be recorded and then only in the coronal plane. Moreover, this technique cannot record the effect on the food produced by the lower molars as they traverse the uppers during the power stroke of a chewing cycle. An occlusal analysis is essential if the exact path of movement in terms of the architecture of the molar crowns is to be documented and the resultant effect on food between the teeth to be inferred.

Living representatives of the primates show considerable variation in the form and functional design of their molar teeth. If evolutionary changes in the type and direction of occlusal contact during the power stroke can be documented, then functional explanations for this variability can be advanced. One should keep in mind, however, that the movements of mastication and the functional designs of the cheek teeth in primates are a reflection of natural selection for specialization in food preference, acting within the limits of mammalian craniofacial architecture and neuromuscular control.

The molars of *Tupaia, Galago, Saimiri,* and *Ateles* have been examined in order to determine the principles of functional design in primate molar teeth. Figure 7 shows the major features on the occlusal surfaces of these teeth. Those reflecting both similarities and differences in their functional relations during the power stroke are discussed and correlated with the behavioural data from the experimental study. If the functional morphology of molars from fossil primates follows the same principles, then the behavioural results from the extant group can be applied to elucidate some of the traits observed in primate dental evolution. This approach has been used, firstly, to evaluate the primate characteristics of the molars of *Palenochtha minor* from the Paleocene of North America as compared with typical mammalian 'tribosphenic' molars from the Upper Cretaceous and, secondly, to compare *P. minor* with *Aegyptopithecus zeuxis*, a specialised Middle or Late Oligocene primate that may well be an early common ancestor for the pongidae and hominidae.

1. Abrasion and Attrition in the Power Stroke

There are, as demonstrated above, two types of power strokes which can be clearly distinguished on behavioural criteria; *puncture-crushing* strokes are characterised by the failure of the teeth to approach the intercuspal range. At the point of minimal gape (maximal approach to occlusion), food held between the teeth on the active side of the jaws is pulped and crushed between the tips of the cusps of the post-canine tooth row. This behaviour produces a distinctive type of wear: the cusps are blunted, the enamel is worn away, and the exposed softer dentine cratered (fig. 8).

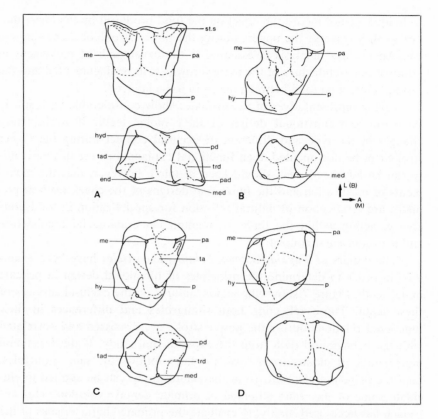

Fig. 7. Occlusal views of upper and lower second molars of *Tupaia glis* (A), *Galago crassicaudatus* (B), *Saimiri sciureus* (C), and *Ateles* sp. (D) (all from the MCZ collection) to illustrate the main features on their crowns. The circular or sub-circular areas at the junctions of solid lines are the cusps; the solid lines indicate crests or ridges on their slopes or connecting the cusps; and the dotted lines indicate the valleys between them. The same convention has been adopted for figure 10. Abbreviations (also used in fig. 10): A = anterior; end=entoconid; hy=hypocone; hyd = hypoconid; hycd = hypoconulid; L = lateral; me=metacone; med=metaconid; mec = metaconule; mes=metastyle; pa=paracone; pad=paraconid; pac=paraconule; pas = parastyle; p = protocone; pd = protoconid; st.s=stylar shelf; ta=talon basin; tad = talonid basin; tr = trigon basin; trd = trigonid basin.

This is the typical appearance of wear resulting from *abrasion*. Later in the masticatory sequence, the food has been softened; and the teeth achieve an intimate relationship in the power stroke. Worn areas appear on the slopes of cusps or ridges as a result of repeated transitory contact

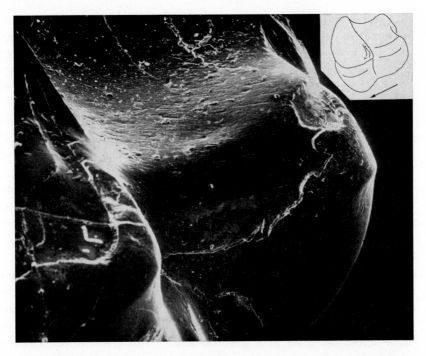

Fig. 8. Scanning electron micrograph of a lower second molar from *Saimiri sciureus* in antero-lateral view. The protoconid is seen in the left foreground, the hypoconid in the right background. Both cusps show the loss of enamel and cavitation of the dentine resulting from *abrasion* in puncture-crushing. The slopes of the crests show striations from *attrition* in chewing. The striations on the floor of the talonid basin are continuous with those on the posterior face of the trigonid (the protocristid) and demonstrate the smooth and uninterrupted transition from phase I into phase II of the power stroke.

between the teeth or between the teeth through a thin film of food and associated material. Examples of wear surfaces formed during this process of *attrition* are shown in figure 8. As can be seen in the scanning electron micrograph, these surfaces are covered by striations or grooves orientated sub-parallel to the direction of relative tooth movement. During chewing power strokes, the direction of movement between contacting teeth is governed by their cuspal patterns, which 'guide' the lower molars in their traverse of the uppers.

Chewing power strokes in the living animals studied are characterised by two behaviourally distinct and sequential phases. During phase I, the lower molar teeth, following their initial contact with the corresponding upper molars, move upwards and antero-medially into centric occlusion. The lower molars then continue to move antero-medially but slightly downwards across the upper molar crowns on the active side during phase II. Since the wear striations are formed during both phases, this smooth change in direction during the power stroke can be traced.

Primate molars are designed to shear, crush, and grind the food interspersed between them in the power stroke. Design features related to each of these modes of food preparation are discussed fully in KAY and HIIEMÄE [1973b]. Briefly, food is sheared between the leading edges of wear surfaces that are orientated so that maximum force may be applied in a direction parallel to the plane of contact between them. Upper and lower shearing blades are reciprocally curved so that a lozenge-shaped space is formed between them on contact, and this diminishes in size as the lower blade moves past the upper. Alternatively, the food may be crushed between occlusal surfaces as a whole, as in puncture-crushing, or between the wear surfaces themselves by muscular forces directed normal to the plane of occlusal contact. Finally, there are matching wear surfaces formed on molars that are produced by the relative movement of one plane across the other in a grinding action. This effect results from the application of forces that have large components of force both normal and parallel to the occlusal plane. It has been argued [CROMPTON and HIIEMÄE, 1970] that a grinding action cannot occur in a steep downward and antero-medially directed movement out of centric occlusion. This may well be correct, if the action is regarded as contingent on the application of substantial compressive forces. However, since the grinding effect on the food can result from simple 'drag' between two surfaces held rather than forced together, the term 'grind' has recently been redefined [KAY and HIIEMÄE, 1973b] to cover both situations: the near-vertical movement found in primitive molars, where 'drag' may be the grinding mechanism; and the near-horizontal movement found in more advanced forms, where the surfaces are nearly flat (in respect of the occlusal plane) and substantial compressive forces are certainly involved.

2. Occlusal Relations in the Power Stroke

Given the relationships between tooth form and mechanism of action discussed above, it is possible to apply these interpretations to the morphological differences between the living animals used in this study. The durations of the preparatory, power, and recovery strokes of the masticatory cycle have been shown to be constant for both puncture-crushing and for chewing; but the power stroke itself has the same relative duration in both types of masticatory behaviour. Although this movement takes the same relative amount of time, irrespective of the effect on the food, there are interesting differences in the length and direction of that movement in the four animals studied. HIIEMÄE and CROMPTON [1971], using the two techniques employed in this study, showed that in the American opossum, *Didelphis marsupialis,* the power stroke consists of a simple upward, medial, and slightly anterior movement from the lateral (buccal) position to centric occlusion. At this point, the stroke was effectively complete, and the lower jaw was then depressed nearly vertically in the recovery stroke. This report shows that in highly evolved primates such as *Saimiri* and *Ateles* the power stroke is not completed at centric occlusion but is continued *antero-medially* in the intercuspal range before the recovery stroke begins. In *Tupaia,* a primate-like insectivore (or a primitive and specialised primate), there is some cinefluorographic evidence that suggests that the lower jaw may move very slightly past centric occlusion before the beginning of the power stroke. Such a movement occurs in *Galago.* These observations suggest that there has been a positive trend towards utilizing the portion of the upper molar lingual to the centric position (fig. 9) and that this development has been achieved by the incorporation of the first part of the downwardly-directed recovery stroke into the power stroke. Once this trend had begun, the new component could be further expanded and its effectiveness enhanced by modifications in molar form associated with the development of larger grinding areas. It is, therefore, important to determine whether such morphological changes have taken place. The living species used in this study can be arranged in a morphological series from 'primitive' to 'advanced' that reflects the observed behavioural transition in the power stroke.

Schematic occlusal views of the upper and lower second molars of *Tupaia, Galago, Saimiri,* and *Ateles* are shown in figure 9. The position of the hypoconid at the beginning of phase I is shown by the base of the upper arrow, at its completion (centric occlusion) by a hollow circle, and at the completion of phase II by the tip of the lower arrow. The movement path between these three points is indicated by the direction of the arrows, and its length by their length. If the length of phase I is expressed as a ratio of the length of phase II, the relative importance of the two phases is demonstrated. The ratio for *Tupaia* is 1.3:1, for *Galago* 0.7:1, for *Saimiri* and *Ateles* 0.5:1. The explanation for these demonstrable pro-

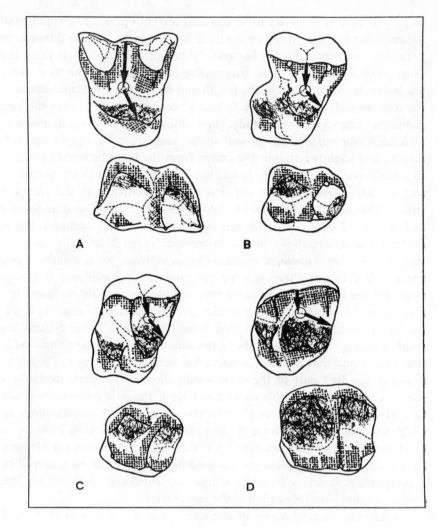

Fig. 9. Occlusal view of the same teeth as shown in figure 7, but with the areas associated with phase I shearing-contact shaded in a regular pattern and those acting as crushing surfaces at the end of phase I and as grinding surfaces during phase II irregularly shaded. The upper arrow shows the path of the lower hypoconid across the upper molar in phase I to centric occlusion (hollow circle) and the lower arrow, the antero-medial movement in phase II. The relative lengths of the arrows indicate the relative lengths of the two phases.

portional differences is illustrated schematically in the figure, where shearing wear facets are shown as regularly shaded and grinding facets are shown as randomly hatched. The former correspond to facets created during phase I contact, the latter to facets traversed in phase II.

The leading edges of phase I wear facets primarily have a shearing function; *Tupaia* and *Galago* illustrate two variants. In the former, the leading edges are concave and on occlusal contact are seen to be reciprocally curved so that, on moving past each other, they form a diminishing 'lozenge-shaped' space which traps the food; in *Galago*, a single lower molar shearing blade may cross the leading edges of two upper molar facets. A longer phase I is almost invariably associated with a predominance of shearing. Phase I is very short in *Ateles*; and, although shearing edges are present, all contact occurs almost simultaneously, in marked contrast to the sequence of events in *Tupaia* or *Galago*. Moreover, the shearing edges on both upper and lower molars of *Ateles* surround the basins, a condition less often seen in *Tupaia*. The arrangement in *Ateles*, a system of basins ringed by sharp-edged blades, is analogous to the type of vegetable cutter where peripheral blades cut the material while it is under compression rather than to a mincing machine, as suggested by MILLS [this volume, pp. 71–72].

The randomly 'scratched' areas in figure 9 are surfaces which approach or actually contact at the end of phase I, crushing the food between them. The same surfaces act as grinding areas in phase II. As can be seen in the figure, *Ateles* has much more surface area in this category than *Tupaia*. This difference is reflected not only in the expansion of the crushing areas on the protocone and in the talonid basin, but also by the presence in *Ateles* of an accessory crushing system between the hypocone and the trigonid basin.

There is, therefore, a very close correspondence between the power stroke as seen cinefluorographically and its phases, direction, and effect in terms of molar tooth design. Comparable design features found in living and fossil primate molars suggest comparable behaviour in the power stroke of mastication.

3. The Power Stroke in Fossil Primates

The molars of *P. minor* are shown in figure 10. This species was described by GIDLEY [1923] and later by SIMPSON [1937], who wrote that this genus was 'the least aberrant of the present genera (of Paleocene primates) as regards comparison with an abstract protoprimate dentition'.

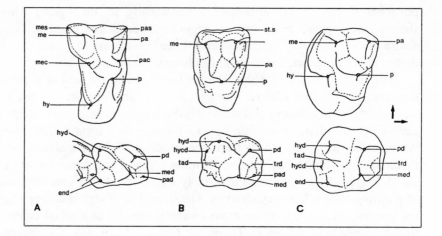

Fig. 10. Occlusal views of (A) the upper and lower first molars, with part of the lower second molar, of *Palenochtha minor* (AMNH 35443 and AMNH 35450, respectively); (B) the upper and lower second molars of *Pelycodus trigonodus* (YPM 25012 and YPM 23317, respectively); and (C) the upper and lower first molars of *Aegyptopithecus zeuxis* (YPM 23798 and YPM 21032, respectively) to show the major features on their crowns. The abbreviations used and the conventions adopted are the same as in figure 7.

Another genus, *Purgatorius* [VAN VALEN and SLOAN, 1965], may be more primitive by this standard of comparison; but it is only known from isolated teeth. *P. minor* shows an important combination of primitive functional traits as well as adaptations foreshadowing later primate specialisations.

The upper molars of *P. minor* are nearly triangular and elongated medio-laterally (bucco-lingually). There are three principle cusps, two small stylar cusps, two conules, and a very small and indistinct hypocone (fig. 10). The stylar shelf is very reduced and without accessory stylar cusps or crests. The lower molars have high trigonids and small talonid basins. The paraconid is present but reduced in size; the hypoconulid is also small. Phase I of the power stroke results in the formation of seven wear facets on the second lower molar; the leading edges of these facets shear upwards and antero-medially past a staggered sequence of upper wear facets. Each shearing surface is reciprocally curved, and several are strengthened by terminal buttressing cusps. Wear surfaces associated with whatever crushing areas are present in the trigonid and talonid basins (shown as 'scratched' in the figure) come into contact with the lateral and postero-lateral surfaces of the protocone as the first phase of the power stroke is completed. The shearing system of staggered blades, producing '*en echelon*' shear [KAY and HIIEMÄE, 1973b] in the upper molars, is topographically quite distinct from the crushing areas, although the latter are partially bordered by the blades. During phase II, the talonid and trigonid basins are carried antero-medially and downwards across the lateral and postero-lateral surfaces of the protocone, thus grinding any food present. The small crushing areas utilised in phase I have a grinding action in phase II.

The modern functional analogue for this system is broadly represented by *Tupaia* although the extreme size of the stylar shelf, the absence of a postprotocrista, and the interlocking relationship between the large paraconid and the hypoconulid in the tree-shrew are like those of no established primate, living or fossil. The dental evidence, however, does not, save on purely morphological criteria, lend support to any argument as to the taxonomic status of the tree shrew.

A detailed comparison between *P. minor* and Late Cretaceous mammals previously suggested to be related to the ancestry of primates will not be attempted here. CROMPTON [1971] has discussed the functional aspects of the emergence of 'tribosphenic' molars. Several of the features that distinguish such teeth from those of *Palenochtha* show that specialisations common to and characteristic of the Primates were developing by the Paleocene. The evolution of tribosphenic molars was characterised by the enhancement of shearing design features effective in phase I; transverse movement was accentuated and shear utilised. The addition of the shearing blades medial to the stylar shelf further enhanced this mechanism. Striations in the basins of the molars of *Gypsonictops* and *Didelphodus* suggest that phase II, in the sense described above, was developing by the Upper Cretaceous, although at that stage it was still functionally insignificant.

Progressive behavioural adaptations which characterize the molars of *Palenochtha* include the diminution of the primitive shearing crests on the stylar shelf. This can be interpreted as a movement away from an extreme form of *en echelon* shear. One wear facet [No. 5; see KAY and HIIEMÄE, 1973b] on the upper molars incorporated in the *en echelon* shearing system of tribosphenic molars has come to border the talonid basin of *P. minor*. The basin itself, although relatively large by comparison with that of tribosphenic molars, is much smaller than in later primates. The relative reduction in the slope of the medial surface of the hypoconid results in a

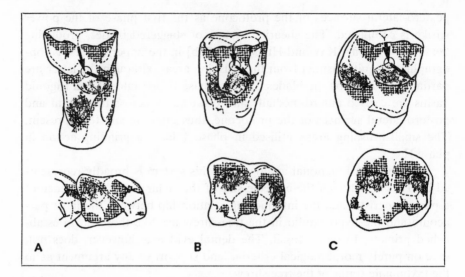

Fig. 11. Occlusal view of the same teeth as in figure 10 but with the phase I and phase II wear facets shaded as in figure 9. Comparison of this figure with figure 9 illustrates the same trends in molar evolution in both the extant and fossil series, namely the relative enhancement of phase II with a shift from *en echelon* shear in the 'primitive' members of each series to compartmentalising shear in the more 'advanced' forms, *Ateles* and *Aegyptopithecus.*

reduction of the angle between it and the occlusal plane, thus making phase II a significant part of the chewing power stroke.

There is some diversity in the molar form of archaic Paleocene primates, indicating a small behavioural radiation. Many subsequent adaptive radiations of primates have occurred that can be identified from the known fossils and the present zoogeography of the order. A behavioural trend among certain primate lineages has been toward specialisations, as judged from molar form, for a diet of vegetable matter with a high intracellular water content and weak cellular structure. Fruit eaters fall into this category. *Ateles* has this masticatory specialisation, as does *A. zeuxis* from the Middle or Late Oligocene of Egypt (fig. 10). The path of movement of the lower molar hypoconid across the crown of the upper molar in the power stroke is shown in figure 11. The distance traversed by the hypoconid is short in phase I when compared with *P. minor* (cf. fig. 9). The shearing blades that contact in phase I (shown with regular stipple in fig. 11) are short in *A. zeuxis,* and those on the lateral (buccal) surface of the upper molars are antero-posteriorly (mesio-distally) directed. There is simultaneous contact between the leading edges of all phase I wear facets. This has been achieved by a shift in the proportions of the crown and the elimination of the primitive *en echelon* shearing system. The scratched areas in figure 11 are those that contact at the end of phase I. As compared with *P. minor* and *Pelycodus* sp., an intermediate stage, these areas are large, the result of expansion of both the talonid basin and the protocone and of the addition of new crushing areas between the trigonid and the hypocone. The angle between these surfaces and the occlusal plane is much flatter than in *Palenochtha,* so that the force of the jaw musculature will be directed nearly normal to the occlusal plane, thus enhancing the crushing and grinding action of the surfaces. The shearing edges used in phase I surround the crushing basins in *A. zeuxis* so that the simultaneous contact of the blades will cut a portion of the food, which is then compressed into the basin and subsequently ground. This mechanism for *compartmentalising shear* is highly developed in *Aegyptopithecus* and has resulted from a gradual change in the proportions of the tooth and of the components of the power stroke.

General Discussion

The observations presented in this paper justify the drawing of four major conclusions concerning basic trends in the evolution of primate mastication.

First, there has been a definite shift from *ingestion-by-mastication* as a function of the post-canine teeth to *incisal biting,* accompanied by the evolution of the spatulate incisors characteristic of the higher primates. The fusion of the mandibular symphysis may also be linked with this transition.

Second, the primates show two distinct types of masticatory behaviour, reflected in the *puncture-crushing* and *chewing* cycles. Further, the path of movement characteristic of each cycle has not been significantly affected by major morphological changes, including the fusion of the mandibular symphysis, the relative shortening of the tooth row, and the relative elevation of the mandibular joint above the occlusal plane.

Third, the timing of the various components of each cycle type has remained more or less constant so that the relative durations of the preparatory and recovery strokes are characteristic for puncture-crushing and chewing cycles. The relative duration of the power stroke is independent of the type of cycle and has not been altered by the changes in the occlusal form of the molars; therefore, the effect on the food undergoing triturition is unchanged.

Fourth, there has been a change from a molar pattern designed for shearing or cutting the food with only very minor and coincidental grinding capacity to a molar pattern designed to cut, crush, and grind the food even after its initial pulping. With this change, there has been an incorporation of the first part of the recovery stroke into the power stroke so that the latter has an initial phase I involving lower molar excursion from the lateral (buccal) position to centric occlusion and a terminal phase II from the centric position to occlusal clearance.

It therefore appears reasonable to infer, since the morphological changes seen in the living series are paralleled by the fossil forms described, that a behavioural pattern for mastication evolved early in mammalian history and that the primates have evolved the structural components of their jaw apparatus within the physiological limits dictated by this behaviour pattern.

Acknowledgements

In addition to our appreciation of the help given to us by Drs. GARCIA and TRUM of the New England Regional Primate Research Center, we should like to express our indebtedness to Dr. R. W. BAKER for giving much of his time and expertise to the statistical problems posed by our results; to Drs. A. W. CROMPTON, D. R. PILBEAM, and M. R. ZINGESER for their encouragement; to Mr. P. BARNETT for photography; and last, but by no means least, to Miss DENISE DONALD for graciously giving up much of her spare time to typing the manuscript.

References

AHLGREN, J.: Mechanisms of mastication. Acta odont. scand. *24:* suppl. 44: 5–109 (1966).

BEYRON, A.: Occlusal relations and mastication in Australian aborigines. Acta odont. scand. *22:* 597–678 (1964).

BUTLER, P. M.: The milk molars of Perissodactyla, with remarks on molar occlusion. Proc. zool. Soc., Lond. *121:* 777–817 (1952).

BUTLER, P. M. and MILLS, J. R. E.: A contribution to the odontology of *Oreopithecus.* Bull. brit. Mus. natur. Hist. Geol. *4:* 1–26 (1959).

CHARLES-DOMINIQUE, P.: Eco-ethologie des prosimiens du Gabon. Rev. Biol. gabon. *VII:* 121–228 (1971).

CROMPTON, A. W.: Studying function by X-ray. Discovery (Peabody Mus. natur. Hist.) *3:* 50–51 (1968).

CROMPTON, A. W.: The origin of the tribosphenic molar in early mammals; KERMACK and KERMACK Early mammals, pp. 65–87 (Academic Press, London 1971); reprinted as suppl. 1, J. Linn. Soc. (Zool.) *50* (1971).

CROMPTON, A. W. and HIIEMÄE, K. M.: Molar occlusion and mandibular movements during occlusion in the American opossum, *Didelphis marsupialis.* J. Linn. Soc. (Zool.) *49:* 21–47 (1970).

FRISCH, J. E.: Trends in the evolution of the hominoid dentition. Bibl. primat., vol. 3 (Karger, Basel 1965).

GIDLEY, J. W.: Paleocene primates of the Fort Union, with discussion of the relationships of Eocene primates. Proc. U.S. nat. Mus. *63:* 1–38 (1923).

GINGERICH, P. D.: Functional significance of mandibular translation in vertebrate jaw mechanics. Postilla *152:* 1–10 (1971).

GINGERICH, P. D.: Molar occlusion and jaw mechanics of the Eocene primate *Adapis.* Amer. J. phys. Anthrop. *36:* 359–368 (1972).

HIIEMÄE, K. M.: Masticatory function in mammals. J. dent. Res. *46:* 883–893 (1967).

HIIEMÄE, K. M. and ARDRAN, G. M.: A cineradiographic study of feeding in *Rattus norvegicus.* J. zool. Soc., Lond. *154:* 139–154 (1968).

HIIEMÄE, K. M. and CROMPTON, A. W.: A cinefluorographic study of feeding in the American opossum, *Didelphis marsupialis*; in DAHLBERG Dental morphology and evolution, pp. 299–334 (University of Chicago Press, Chicago 1971).

HIIEMÄE, K. M. and KAY, R. F.: Trends in the evolution of primate mastication. Nature, Lond. *240:* 486–487 (1972).

HLÀDIK, A. et HLÀDIK, C. M.: Rapports trophiques entre végétation et primates dans la forêt de Barro-Colorado (Panama). Terre Vie *I:* 25–117 (1969).

KAY, R. F. and HIIEMÄE, K. M.: Mastication in *Galago crassicaudatus:* a cinefluorographic and occlusal study; in MARTIN Proc. prosimian Biology Research Seminar, London 1972 (Duckworth, London, in press, 1973a).

KAY, R. F. and HIIEMÄE, K. M.: Jaw movement and tooth use in recent and fossil primates. Amer. J. phys. Anthrop. (in press, 1973b).

LAHEE, F. H.: Field geology; 6th ed. (McGraw Hill, New York 1957).

MILLS, J. R. E.: Ideal dental occlusion in the primates. Dent. Practit. *6:* 47–61 (1955).

MILLS, J. R. E.: Occlusion and malocclusion in the teeth of primates; in BROTHWELL Dental anthropology (Oxford University Press, Oxford 1963).

MILLS, J. R. E.: A comparison of lateral jaw movement in some mammals from wear facets on the teeth. Arch. oral Biol. *12:* 645–661 (1967).

MILLS, J. R. E.: Evolution of mastication in primates; in ZINGESER Craniofacial biology of primates, vol. 3: Symp. Proc. 4th int. Congr. Primat., Portland, Ore., 1972 pp. 65–81 (Karger, Basel 1973).

MØLLER, E.: The chewing apparatus: an electromyographic study of the action of the muscles of mastication and its correlation with facial morphology. Acta physiol. scand. *69:* suppl. 280 (1967).

SIMPSON, G. G.: The Fort Union of the Crazy Mountain Field, Montana, and its mammalian faunas. Bull. U.S. nat. Mus. *169:* 1–287 (1937).

SZALAY, F. S.: Mixodectidae, Microsyopidae and the insectivore-primate transition. Bull. amer. Mus. natur. Hist. *140:* 193–330 (1969).

VALEN, L. VAN: Deltatheridia, a new order of mammals. Bull. amer. Mus. natur. Hist. *132:* 1–126 (1966).

VALEN, L. VAN and SLOAN, R. E.: The earliest primates. Science *150:* 743–745 (1965).

Authors' addresses: Dr. KAREN M. HIIEMÄE, Unit of Anatomy with Special Relation to Dentistry, Guy's Hospital Medical School, *London S.E. 1* (England); RICHARD F. KAY, Division of Vertebrate Paleontology, Peabody Museum, Yale University, *New Haven, CT 06520* (USA)

Symp. IVth Int. Congr. Primat., vol. 3: Craniofacial Biology of Primates,
pp. 65–81 (Karger, Basel 1973)

Evolution of Mastication in Primates

J. R. E. MILLS

Institute of Dental Surgery and University College, University of London

Introduction

Surprisingly, little is known about the functioning of the dentition in
the mammal-like reptiles, but there can be little doubt that in the earlier
members the upper teeth overhung the lower and did not at any stage
meet them to produce chewing. The lower jaw was capable only of a vertical snapping movement produced by action of the reptilian adductor and
pterygoid muscles, so that prey could be grasped by the teeth to prevent
its escape. It would then be swallowed whole or torn apart by a number
of competing animals, as happens with recent reptiles. Chewing, that is
the use of the teeth to break up the food into more manageable morsels,
was undoubtedly present in some groups of late mammal-like reptiles
[CROMPTON and JENKINS, 1968; PARRINGTON, 1967]; but these forms were
herbivores, and jaw movement was apparently propalinal. In the carnivorous group that led to recent therian mammals, however, the jaw movement was still vertical and tooth function confined to seizing and tearing
the prey.

The use of the teeth for chewing is, therefore, essentially a mammalian characteristic, necessary to improve the efficiency of feeding in the
very small animals which were the first mammals in order to provide for
the increased metabolism which came with warm blood. The purpose of
the present paper is to trace the development of this chewing from a
primitive cutting action of the earliest mammals to the elaborate grinding
mechanism of the higher primates.

In the descriptions which follow, I shall use the well-known COPE-
OSBORN terminology for the cusps of the posterior teeth; while crests

will be indicated according to the terminology of SZALAY [1969]. The surfaces of the teeth will be indicated by standard dental terminology. The buccal surface is the outer surface, toward the cheeks; the lingual is the inner, toward the tongue. The mesial surface is that nearest to the midline, as one moves around the arch of teeth; that is, for the molar teeth it is anterior. The distal is the opposite side, the posterior of a molar. In describing upper teeth, I have assumed the jaw to be positioned with the teeth uppermost, so that terms such as 'above' indicate toward the tips of the cusps, while 'below' means toward the gum margin.

Mesozoic Mammals

The earliest known therian mammal, which may well lie on or close to the main line of evolution to recent placental and marsupial mammals, is *Kuehneotherium* from the uppermost Triassic of South Wales. This animal apparently had both reptilian and mammalian jaw joints functioning simultaneously. It is known only from a fairly large collection of isolated teeth and a few fragments of dentaries and maxillae and has been described by KERMACK *et al.* [1965, 1968]. The molar teeth carry three main cusps, arranged when viewed from above (fig. 1A) in the form of an obtuse-angled triangle. In the upper jaw, the angles of the triangle are formed by the paracone, stylocone, and metacone, and in the lower by the protoconid, paraconid, and metaconid. The triangles so formed are not of the isosceles variety, the mesial side of the upper and distal of the lower being more transversely arranged than their opposite sides. The angle that these sides subtend at their apex (the paracone or protoconid) varies considerably between individual teeth, probably according to their position in the tooth row.

The crests that run mesially and distally from the tips of these cusps are sharp, and chewing would seem to use these crests to cut the food in a manner analogous to the blades of a pair of scissors. The buccal surface of the lower tooth is pressed firmly against the lingual surface of the upper tooth (fig. 1A, 2A) without food lying between these surfaces. The morsel of food lies on the upper surface of the lower tooth, where it may be impaled on the pointed cusps; and as the jaw is closed, the sharp crests cut it. This process will only be efficient if the teeth are closely pressed together; a pair of scissors with a loose joint is useless. To enable the animal to press the teeth together, and at the same time to

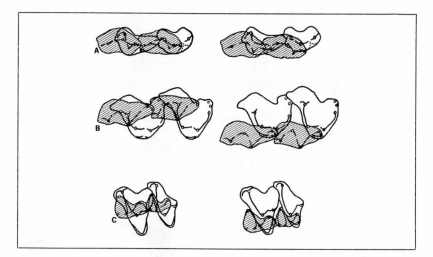

Fig. 1. Upper and lower molar teeth of Mesozoic mammals superimposed *left* about the beginning of the cutting stroke, *right* at the end of the stroke, centre in centric relationship. A *Kuehneotherium* (occlusion assembled from isolated teeth). B *Peramus tenuirostris* [based on information from CLEMENS and MILLS, 1971]. C upper teeth of *Pappotherium pattersoni* superimposed on probably congeneric 'Type 6' molar; the posterior lower molar is hypothetical.

release the pressure when required, a degree of lateral jaw movement is essential in addition to vertical movement.

A somewhat later stage of dental development is represented by *Peramus* (fig. 1B, 2B) from the uppermost Jurassic. The known specimens of this animal have recently been reviewed by CLEMENS and MILLS [1971], who believe that, although reasonably representative of this stage of therian evolution, it lies somewhat off the main line thereof. Here again, we have the principle of reversed triangles, with the lower triangles shearing between two upper ones and vice versa (fig. 1B, 2B). Once again, the action is a cutting one, akin to that of a pair of scissors. The buccal surfaces of the lower teeth are pressed closely against the lingual surface of the upper, without the interposition of food. The morsel of food is placed on the upper surface of the lower teeth and kept in position by upper and lower cusps, while the sharp ridges cut it into pieces suitable for ingestion. The triangular outline of the teeth is accentuated as compared with even the most acute-angled *Kuehneotherium* molars. If the cusps and their asso-

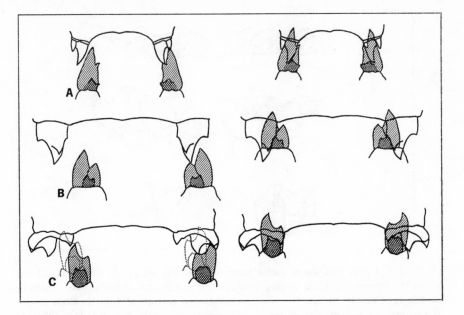

Fig. 2. Hypothetical cross-section through the jaws in the molar region of: (A) *Kuehneortherium praecursoris,* (B) *Peramus tenuirostris,* and (C) *Pappotherium pattersoni; left* at the beginning of the cutting stroke, *right* at the end of the cutting stroke in centric occlusion. The talonids of the lower molars are cross-hatched.

ciated crests were arranged in sraight lines mesio-distally, then, as the jaw closed, all points on the cutting edges would come into action virtually simultaneously. By having the crests arranged somewhat transversely, as indicated in figure 2, only one point on a crest is in contact with its opponent at a given time, as in a pair of scissors; and all the force of the bite is concentrated at this one point, with much increased efficiency.

CROMPTON and HIIEMÄE [1970], from cineradiographic studies of the American opossum, *Didelphis marsupialis,* have shown that during chewing in this animal the whole mandible, including the condyle, moves medially and, to some extent, anteriorly. I have suggested [MILLS, 1967a], from a study of the wear on the teeth of a number of recent animals with primitive types of dentition, that this is the original type of jaw movement in the earliest therian mammals. The asymmetrical outlines of the trigonids and trigons would seem to confirm this suggestion; the direction of movements would probably bisect the angle formed by the two crests that meet at the protoconid of the lower molar and at the lingual apex of the upper. CROMPTON and HIIEMÄE have also shown in *Didelphis* that the mandibular symphysis is very mobile, and this mobility again would seem to be the primitive condition in mammals. It was apparently the case in *Peramus* [CLEMENS and MILLS, 1971].

During chewing in primitive mammals the jaw is closed by the main temporalis-masseter adductor muscle. Simultaneously, an inward and slightly forward translatory movement is necessary. Such a movement presents problems because of the mobile symphysis and the consequent tendency for the mandibular body to rotate about an antero-posterior axis. To be efficient, it must move in a parallel manner or with, at most, slight and controlled rotation. The muscles responsible for this component of jaw movement can be deduced from the work of HIIEMÄE and JENKINS [1969] on *Didelphis*. It would seem that the medial movement is brought about by the internal and external pterygoids. The internal pterygoid is probably inserted close to the lower border of the mandible, from the tip of the angular process forwards to a point roughly level with the anterior border of the ascending ramus. This long insertion would give control against rotation about a vertical axis, since the angular process extends backwards behind the condyle. A second muscle, the external pterygoid, is inserted into the head of the mandibular condyle. The combined effect of the inward pull of the external pterygoid at the upper extremity of the mandible and the internal pterygoid at the lower gives rotational control about an antero-posterior axis. The mandible would be drawn laterally, on a reciprocal path, for the beginning of the next stroke by the more horizontal fibres of the masseter (zygomatico-mandibularis), with some posterior direction supplied by the suprazygomatic fibres of temporalis [MILLS, 1967a].

During the first half of known mammalian history, from the late Triassic to the end of the Cretaceous, the mammals remained an insignificant group of small animals in a land dominated by dinosaurs. It is therefore not surprising that they are known from very few specimens. By the middle of the Cretaceous, SLAUGHTER [1965, 1971] has suggested that the therians had divided into marsupials and placentals. A representative of the latter is shown in figures 1C and 2C. These diagrams have been produced from the upper molars of *Pappotherium* and the only known lower of a type that SLAUGHTER calls 'Trinity type 6' and says is the most likely to be associated with the upper teeth. The more distal lower molar illustrated in figure 1C is an hypothetical reconstruction.

The upper molars have remained triangular in outline and indeed have become wider, bucco-lingually, relative to their length. The lingual cingulum, which was present in *Kuehneotherium*, has become so wide as to be regarded as an important cusp, the protocone. Consequently, the paracone has been displaced buccally, as has, to a lesser extent, the metacone. In the lower jaw, the main development was in the talonid, which SLAUGHTER [1971] now describes as 'basined'. This basin is, however, deficient on the lingual side, between the entoconid and the posterior face of the trigonid. In occlusion the tip of the protocone overlaps the lingual side of the basin at this point; the protocone does not occlude in the centre of the talonid basin as in most primates. The pestle and mortar action between protocone and talonid is certainly not present at this stage.

Chewing was still of a cutting nature, as shown in figure 2C. As the mouth closes, cutting first occurs between the anterior crest from the paracone (paracrista) and the posterior side of the trigonid (protocristid), and also between the posterior metacone crest (metacrista) and the anterior crest of the trigonid (paracristid). At a slightly later stage, shown by the dotted outline in figure 2C, the cutting edges of the trigonid have moved upwards past these upper crests, and have come to cut against the cingular crests which run buccally from the tip of the protocone. At the same time, the hypoconid shears down the groove between the paracone and metacone; and a further cutting edge comes into action as the crests that run lingually from the tip of the hypoconid shear against the centrocrista of the upper molar (fig. 1C). The cutting action is, therefore, at this stage much more sophisticated than in *Peramus,* although not different in principle. The cutting mechanism is duplicated vertically, with the ectoloph region coming into action before the protocone and its crests. It is also, to some extent, duplicated mesio-distally, with incipient development of a second triangular shear, as the hypoconid and its crests shear between the paracone and metacone of the upper tooth. This latter is a development that is not new in *Pappotherium.* A similar, but less well-developed, mechanisms exists for the hypoconid of *Peramus* and even for the middle Jurassic *Amphitherium.* Indeed, a groove exists between the paracone and metacone of *Kuehneotherium,* although in occlusion it is the paraconid that shears down it, since the talonid is at this stage vestigial. I have suggested elsewhere [MILLS, 1967b] the mechanism by which the hypoconid first reinforces and then replaces the paraconid of the succeeding lower molar in this function.

It would seem that there is a relationship between the protocone and a basined talonid, in that one does not exist without the other [BUTLER, 1961; MILLS, 1964]. In many mammals, notably the primates, the protocone occludes in the centre of the talonid in a manner often described as analogous to a pestle in a mortar. CROMPTON and HIIEMÄE [1970] have shown that this condition is not the case in the primitive tribosphenic dentition of *Didelphis*, nor presumably, therefore, in somewhat similar primitive dentitions, including *Pappotherium*. They suggest that the function of this combination of cusps is to use the talonid as an anvil, on which the food is pulverised and pierced by the opposing protocone. This motion doubtless occurs, but I would suggest that an additional and important function for these structures is to provide the sharp cutting edges to give a more efficient cutting mechanism.

It would seem difficult to deduce the method of jaw movement from a few fossil teeth of a long-dead animal. Nevertheless, it seems probable that lateral jaw movement was still purely translatory, as in *Didelphis* [CROMPTON and HIIEMÄE, 1970], with the ipsilateral condyle being drawn medially, and perhaps slightly anteriorly, by the pterygoid muscles. The symphysis was probably still mobile. From the increased width of the teeth, relative to their length, it appears that the lateral component of jaw movement was still further increased.

The Primates

The primates are believed to have evolved from a generalised insectivore in the later part of the Cretaceous, in a group that took up an arboreal mode of life. This change involved a change in diet, and with it fairly major changes in the chewing apparatus. The cutting mechanism that had reached a high pitch of efficiency in *Pappotherium* and its allies is excellent for cutting up the brittle, chitinous bodies of insects and small crustaceans. It is still found among recent small insectivorous mammals. In the trees the diet became more generalised and, to judge from even the most primitive recent primates, included fruits, seeds, and perhaps leaves, as well as small animals, including insects. Chewing gradually changed to a process in which a ridge on one tooth grinds down a groove on the opposing tooth, with the food interposed between the two. I have likened the cutting action of the Mesozoic mammals to a pair of scissors. It is more difficult to find an analogy for the mechanism that now evolved;

but one can be found, not surprisingly, in a domestic mincing machine. In this, the material to be minced is fed in at the top. It is then drawn along by a screw mechanism that is driven by hand or by an electric motor. This screw is a loose fit within the grooved body, and the effect of the 'thread' – of the ridges on the screw within the grooves – is not only to draw the material forward, but also to mince it. This same mincing effect is produced by the ridges of primate teeth shearing down the opposing grooves, with the food in between.

In the earliest type of dentition, the cutting edge was all-important. For example, the cutting edges that run mesially and distally from the hypoconid cut against those which run mesially and distally from the mesostyle to the paracone and metacone, respectively. The near-vertical ridge running down to the gum margin from the tip of the hypoconid shears down the near-vertical groove running down from the mesostyle to the centre of the upper molar. At that stage, this cusp-in-groove action had no special function in chewing; it was merely the point at which the cutting edges changed direction. With the change to the primate type of chewing, this ridge-in-groove action becomes all important; and as it evolves, jaw movement becomes more horizontal, as do the ridges and the grooves in which they work. The significance of the cutting edges disappears, and the cusps become more rounded and blunt. The cusps of the upper molars effectively rotate about an axis through the parastyle, mesostyle, and metastyle, so that the tips of the paracone and metacone are displaced buccally, while the styles themselves sink into oblivion. A parallel change took place on the lower teeth, and is partly responsible for the disappearance of the paraconid.

This change may be illustrated by considering the three primates illustrated in figure 3. *Purgatorius ceratops* from the uppermost Cretaceous may well be the oldest known primate, although known only from an isolated lower molar. The related *Purgatorius unio* is somewhat later, from the early Paleocene [VAN VALEN and SLOAN, 1965; SZALAY, 1969]. At first sight its molar teeth are a logical development of *Pappotherium* and its relatives from the Trinity sands. The upper molar is still basically triangular; but the buccal row of styles is much reduced, with the paracone and metacone displaced toward the buccal margin of the tooth. The protocone has come to lie within the talonid of the lower molar, which is now completely basined. On the lower molar, the talonid is now somewhat wider bucco-lingually than the trigonid, indicating its increased importance in function. The protocone working within this basin may already

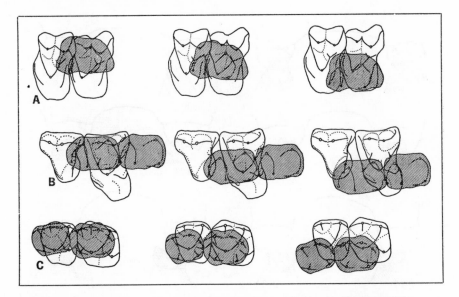

Fig. 3. Upper and lower molar teeth of fossil and recent primates superimposed *left* about the beginnning of the chewing stroke, *centre* at the moment when centric occlusion is reached, and *right* at the end of the lingual phase of occlusion. A *Purgatorius unio* assembled from isolated teeth [from VAN VALEN, 1965, and SZALAY, 1969]. B P⁴, M¹ of *Palenochtha* on M_1, M_2 of *Mckennatherium;* C *Gorilla.*

have started the pestle and mortar action that is the typical chewing action of later primates.

The main chewing action here is intermediate between the pure cutting of the Mesozoic mammals and the pure grinding or mincing of later primates. The sides of the trigon and trigonid have become much less vertical; and it seems probable that the food was milled between these surfaces, and not merely cut by their edges. Mesial displacement of the protocone and distal displacement of the metaconid have produced a preprotocrista and protocristid that are unusually straight. An analogous situation exists in the recent Erinaceidae [MILLS, 1966], and in that family the principal shear is between the surfaces adjacent to these crests.

It is probable that at this stage another important development took place. In the mammals described so far, chewing commenced with the mouth open and the mandible displaced to the ipsilateral side. The stroke proceeded with the mandible moving upwards and medially until the centric position was reached, when the stroke was complete. In *Purgatorius* it

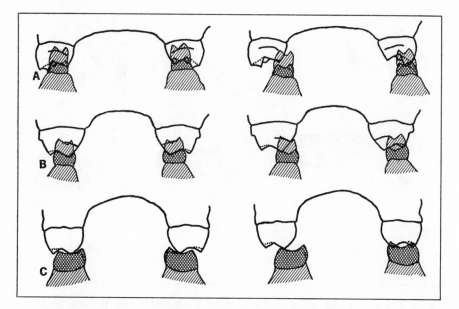

Fig. 4. Cross-sections through the jaws, in the molar regions, of: (A) *Purgatorius unio*; (B) *Palenochtha/Mckennatherium;* (C) *Gorilla,* superimposed *right* at the beginning of the chewing stroke on right side (the left side of the mouth shows more or less the end of the lingual phase of chewing); *left* in centric relationship. A and B are hypothetical reconstructions, but C is based on an actual specimen. The talonids of the lower molars are cross-hatched. Note: in A, the protocone of the upper molar appears superimposed on the entoconid of the lower. In fact, it occludes in a groove mesial to the entoconid. Similarly, the protoconule occludes in a groove mesial to the hypoconid.

seems likely that the stroke continued beyond the centric position, as the mandible continued to move medially but now downwards, into what I have called [MILLS, 1955] the lingual phase of occlusion (fig. 3A, 4A).[1]

If jaw movement is produced by a pure translatory movement of the condyles and we ignore any movement at the symphysis, then as the ipsilateral condyle moves inwards and forwards in the buccal phase of occlusion, the contralateral condyle will move forwards and laterally on a parallel path. During the lingual phase of occlusion on the ipsilateral side, the condyle will move medially and backwards, on a course that is a mirror-

1   This is identical with the phase II of the power stroke of HIIEMÄE and KAY, this volume. [Ed.]

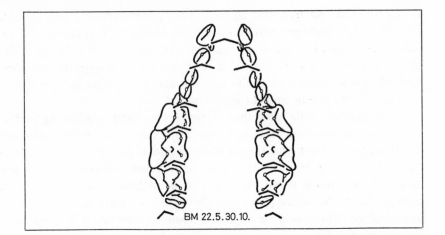

Fig. 5. Outline of the upper dental arch of *Erinaceus*, showing the path of the protoconids of selected lower teeth from the beginning of the chewing stroke on one side of the mouth to the beginning of a chewing stroke on the opposite side. The similar marks behind the dentition, alongside the catalogue number of the specimen, illustrate the movement of the mandibular condyles.

image of that followed during the buccal phase. The contralateral condyle, moving on a parallel course, would then be ready for a chewing stroke on the contralateral side if this were desired. Figure 5 shows the situation in *Erinaceus*. On the outline of the upper teeth are seen a number of thick straight lines in the form of an inverted V. These lines represent the path of the protoconids of selected lower teeth across the upper arch during chewing, apart from the two placed well behind the arch that represent the path of a point within the condyles. These paths have been deduced from a study of the wear on the teeth, as detailed by MILLS [1967a]. It will be seen that the condyle moves forwards and inwards in the buccal phase and backwards and inwards in the lingual, but that the latter is not quite a mirror-image of the former. The only possible explanation would seem to be that as the ipsilateral condyle moves backwards and medially during the lingual phase, the whole mandible rotates in a counter-clockwise direction about a vertical axis through the contralateral condyle, as that condyle moves backwards and laterally. Consequently, the ipsilateral condyle, at the end of the lingual phase, is slightly farther forward than it would be if it were moving on a path parallel to the contralateral condyle. Similarly, during the buccal phase of occlusion, as the ipsilateral condyle moves forward and medially, the mandible rotates about a vertical axis through this condyle in a counter-clockwise direction. This element of rotation also means that the more anterior teeth move farther during a given time than the posterior, which, in turn, move farther than the condyles. It is, of course, additional to the rotation about a horizontal axis that is responsible for opening and closing the mouth.

It is, perhaps, little more than conjecture to apply this finding in *Erinaceus* to the condition seen in *Purgatorius;* but it seems very probable that this type of development took place in lateral jaw movement at about this stage. In the Hominoidea, and many other primates, lateral jaw movement is brought about essentially by a rotation about the ipsilateral condyle during the buccal phase and about the contralateral condyle in the lingual phase. There may be a small amount of translation in some individuals in man [see, e.g., PREISKEL, 1970]; but when present, it is very slight. In *Tarsius* and the Cercopithecoidea (and also in *Tupaia*) there is a combination of translation and rotation in varying degrees [MILLS, 1963]. It would seem, therefore, that lateral jaw movement began with a translatory movement of the condyles equal to the movement of the teeth. As the amount of lateral movement increased during chewing, a situation arose where the glenoid fossa in the squamosal would have to be impossibly wide to accommodate the required movement. This problem was overcome by introducing rotation, which allowed the teeth to move more than the condyles. Some degree of rotation is found in all living primates, so that it probably arose soon after the primates themselves first developed from the basic insectivorous stock.

Another development seen in *Purgatorius* is that of the mesio-lingual and disto-lingual cingula on the upper molars. The latter is rather the larger and is apparently the precursor of the hypocone; while the former is found in some recent primates (e.g., *Lemur*). They would seem to replace the original function of the protocone and its associated cingula, acting as a 'chopping block' for the cutting edges of the trigonid of the lower teeth. It will be seen from figure 3A that the 'hypocone' shears directly against the tip of the paraconid and may be the reason that, as VAN VALEN and SLOAN [1965] point out, this cusp is reduced in height and is not distinct from the rest of the paralophid.

From the later Paleocene, a more definitely primate type of molar pattern is seen in the family Paromomyidae, represented in figure 3B by P^4 and M^1 of *Palenochtha* and M_1, M_2 of the possibly identical *Mckennatherium* [VAN VALEN, 1965]. The principle of reversed triangles is still discernible but is fast being replaced by the grinding type of occlusion, with further development of the lingual phase. The disto-lingual cingulum of M^1 is now definitely a hypocone, although considerably lower than the trigon. The talonid basin is larger in all dimensions than the trigonid, although the latter is still raised somewhat above the former. It would seem clear from the illustrations of VAN VALEN [1965] that there is a well-worn shearing surface across the distal surface of the trigonid that sheared against the mesial surface of the trigon. These surfaces are, however, even less vertical than in *Purgatorius;* and it would seem that the mechanism is one of grinding rather than cutting.

The details of the two phases of occlusion may be seen by comparing the three diagrams in figures 3B and 4B. Starting with the mouth open and the mandible displaced to the ipsilateral side, the mouth begins to close and the mandible to move upwards and medially. The teeth come into contact, with the distal surface of the trigonid shearing across the mesial surface of the trigon (fig. 3B, *left* to *centre*). The distal face of the protoconid shears against the mesial face of the paracristid; while the metaconid, whose distal face is initially against the mesial face of the protoconule, passes lingually against the mesial face of the preprotocristid. The original mesial face of the trigonid is beginning to change its character, with further reduction in height of the paraconid region, which is now level with the posterior margin of the talonid of the more anterior tooth in the entoconid region. The paracristid is therefore tending to be replaced by a new crest that straddles two adjacent teeth, from the protoconid of one molar to the entoconid of its mesial neighbour. The mesial face of this ridge grinds against the distal face of the trigonid, with the entoconid shearing down the shallow groove distal to the tip of the protocone. The distal side of this ridge (which includes the hypoconulid) shears, at least incipiently, against the mesial side of the hypocone. Simultaneously, the hypoconid shears down the groove between the paracone and metacone. This last is the only case where the direction of movement is parallel to the ridge and groove. In the other cases, the direction of shear is at an angle to the ridges involved, a legacy from the cutting occlusion. This motion brings us, momentarily, to the centric relationship shown in figures 3B and 4B, *centre*.

The lower molars then continue, without pause, into the lingual phase. The mandible begins to descend, while continuing to move to the contralateral side. In most recent primates, the hypoconid, during the lingual phase, moves mesially and lingually along the groove between the protocone and protoconule [BUTLER, personal communication],[2] while the protocone moves reciprocally along the groove between the hypoconid and hypoconulid (fig. 3B, *centre* to *right*). When this occurs, the protoconid, raised above the talonid, moves along the groove between the protocone and hypocone, so that the hypocone is necessarily lower than the trigon. The grooves between protocone and protoconule and those between hypocone and protocone must, therefore, be parallel.

That this is the condition in the Paromomyidae again seems to be corroborated by the wear facets on the (lower) molar teeth of *Mckennatherium*, as illustrated by VAN VALEN [1965]. There is a large wear facet extending across the hypoconid and hypoconulid, produced by the buccal and disto-lingual faces of the protocone as it sweeps down the groove between these cusps. There is also a facet on the lingual side of the tip of the protoconid that would be produced by shear against the hypocone.

Assuming this account to be correct, the cusps of the lower molars slide almost directly lingually during the buccal phase, changing to a mesio-lingual direction during the lingual phase. This path would imply a considerable degree of rotation alternately about the two condyles, although it is very probable that translatory movement also occurred, since it is present in the Cercopithecoid monkeys and in *Tarsius*. While the latter might have evolved independently from the Paleocene, it seems unlikely that the Old World monkeys have done so.

The muscles involved in this movement would seem to be as follows: the ipsilateral condyle is drawn forwards and inwards by the pterygoids of the same side, the condyle itself being moved by the external pterygoid. At the same time, the condyle rotates. This action could, conceivably, be produced by the anterior fibres of the temporalis muscle, inserted into the tip of the coronoid process. Such an action would be mechanically very inefficient and, if the symphysis were still mobile, would tilt the mandible inwards. It seems more probable that the rotation results from posterior movement of the contralateral condyle produced by the muscles of that side, notably the suprazygomatic part of temporalis. This latter action would involve comparatively little change of muscle function from that carried out in the previous, purely translatory, lateral jaw movement and would correspond to the action in recent higher primates. It would, however, require a very much stronger symphyseal joint than that found by CROMPTON and HIIEMÄE [1970] in *Didelphis* or than that prob-

2   See also BUTLER, this volume. [Ed.]

ably existing in early mammals. It is notable that in recent Prosimii, the two halves of the mandible are firmly bound together at the symphysis and that this symphysis is long and almost horizontal. In the Anthropoidea, the two halves of the symphysis are fused; and the need for the rotational lateral jaw movement may be the cause.

The story of the development from this stage to the recent Hominoidea is well known. Any vestiges of the original cutting action in chewing were eliminated, and the pestle-and-mortar action became the sole type to be employed. In fact, this term is a misnomer, since, as previously indicated, the action is essentially one of a rather blunt ridge moving down a groove, with food interposed between. The protocone moves down between the metaconid and entoconid in the buccal phase, changing direction to move between the hypoconid and hypoconulid in the lingual phase (fig. 3C). Reciprocally, the hypoconid moves down the groove between paracone and metacone in the buccal phase and mesio-lingually between the protocone and protoconule in the lingual phase, although the protoconule itself tends to be vestigial in recent hominoids.

The success of this type of chewing mechanism is seen in an animal such as *Tetonius* or the recent *Tarsius*, where the size of the talonid came to greatly exceed that of the trigonid. The latter was becoming functionless, with the elimination of cutting in chewing. It returned to usefulness by the development of a replicate of the talonid-trigon action, due to the development of the hypocone. This cusp moves, during the buccal phase, down the groove between the entoconid of one tooth and metaconid of its neighbour; and in the lingual phase it shears across the mesio-lingual face of the protoconid. The protoconid and hypoconulid of the lower molars move on reciprocal paths across the upper tooth, as seen in figure 3C. With the elimination of the cutting action, the paraconid has also disappeared.

The relationship of cusps to grooves during the chewing activity is notably constant in primates, with rather few exceptions of which the most notable is the Cercopithecoid monkeys [MILLS, 1955]. During this development, the extent of the lingual phase of occlusion increases until in the Hominoidea buccal und lingual phases are of equal duration. As the teeth are moving in the buccal phase on one side of the mouth, those on the other side are moving in the opposite direction through the lingual phase (fig. 4). This action produces a balanced occlusion and reduces stress on the jaw joint, although it seems likely that food is chewed on only one side at a time.

Summary

(1) In the carnivorous reptilian ancestors of the mammals, no chewing was possible, the teeth being used to grasp the prey only.

(2) In the earliest mammals, chewing took the form of a lateral movement of the jaw and condyles. This allowed the lower teeth to be pressed against the uppers, so that the sharp crests of the cusps would cut up the hard bodies of the insects that formed a large part of the diet.

(3) With the development of the primates living in trees, a profound change of diet took place. This involved a change in chewing action to one of a more grinding type; and this grinding, in turn, involved a change in dental pattern.

(4) The increased amount of lateral jaw movement could not be accompanied by an equal amount of translatory movement of the condyles, since there was not sufficient space in the glenoid fossa of the squamosal. This difficulty was overcome by introducing rotation of the mandible about a vertical axis through the appropriate condyle. In the higher primates, this rotation has effectively replaced the translatory movement.

(5) This evolution has been illustrated by reference to certain fossil forms, which seem to represent the stage which primate development had reached at the time concerned. It does not follow that these forms lie on the direct line of evolution to the Hominoidea, and indeed it is improbable that they do.

References

BUTLER, P. M.: Relationships between upper and lower molar patterns; in Int. Colloq. on the evolution of lower and non-specialized mammals. Kon. VI., Acad. Wet. Lett. Sch. Kunst. België, part. I, pp. 117–126 (Brussels 1961).

CLEMENS, W. A. and MILLS, J. R. E.: Review of *Peramus tenuirostris* Owen (Eupantotheria, Mammalia). Bull. br. Mus. natur. Hist., Geol. *20:* 89–113 (1971).

CROMPTON, A. W. and HIIEMÄE, K.: Molar occlusion and mandibular movements during occlusion in the American opossum, *Didelphis marsupialis* L. J. Linn. Soc. (Zool.) *49:* 21–47 (1970).

CROMPTON, A. W. and JENKINS, F.: Molar occlusion in late Triassic mammals. Biol. Rev. *43:* 427–458 (1968).

HIIEMÄE, K. and JENKINS, F.: The anatomy and internal architecture of the muscles of mastication of *Didelphis marsupialis*. Postilla *140:* 1–49 (1969).

KERMACK, D. M.; KERMACK, K. A. and MUSSETT, F.: The Welsh pantothere *Kuehneotherium praecursoris*. J. Linn. Soc. (Zool.) *47:* 407–423 (1968).

KERMACK, K. A.; LEES, P. M. and MUSSETT, F.: *Aegialodon dawsoni,* a new trituberculosectorial tooth from the lower Wealden. Proc. roy. Soc. B. *162:* 535–554 (1965).

MILLS, J. R. E.: Ideal dental occlusion in the primates. Dent. Pract. *6:* 47–61 (1955).

MILLS, J. R. E.: Occlusion and malocclusion of the teeth of primates; in BROTHWELL Dental anthropology, pp. 29–51 (Pergamon Press, London 1963).

MILLS, J. R. E.: The dentitions of *Peramus* and *Amphitherium*. Proc. Linn. Soc., Lond. *175:* 117–133 (1964).

MILLS, J. R. E.: The functional occlusion of the teeth of Insectivora. J. Linn. Soc. (Zool.) *47:* 1–25 (1966).

MILLS, J. R. E.: A comparison of lateral jaw movement in some mammals from wear facets on the teeth. Arch. oral Biol. *12:* 645–661 (1967a).

MILLS, J. R. E.: Development of the protocone during the Mesozoic. J. dent. Res. *46:* 883–893 (1967b).

PARRINGTON, F. R.: The origin of mammals. Adv. Sci., Lond. *24:* 165–173 (1967).

PREISKEL, H.: Bennett's movement; a study of lateral mandibular movement. Brit. dent. J. *129:* 372–376 (1970).

SLAUGHTER, B. H.: A therian from the lower Cretaceous (Albian) of Texas. Postilla *93:* 1–18 (1965).

SLAUGHTER, B. H.: Mid-Cretaceous (Albian) therians of the Butler Farm local fauna, Texas; in KERMACK and KERMACK Early mammals. J. Linn. Soc. (Zool.) *50:* suppl. 1, pp., 131–143 (1971) and (Academic Press, London 1971).

SZALAY, F. S.: Mixodectidae, Microsyopidae and the insectivore-primate transition. Bull. amer. Mus. nat. Hist. *140:* 195–330 (1969).

VAN VALEN, L.: A Middle Palaeocene primate. Nature, Lond. *207:* 435–436 (1965).

VAN VALEN, L. and SLOAN, R. E.: The earliest primates. Science *150:* 743–745 (1965).

Author's address: Dr. J. R. E. MILLS, Institute of Dental Surgery, Eastman Dental Hospital, University of London, *London* (England)

Dental Morphology and Systematics

Symp. IVth Int. Congr. Primat., vol. 3: Craniofacial Biology of Primates, pp. 82–100 (Karger, Basel 1973)

Gorilla Dental Sexual Dimorphism and Early Hominid Taxonomy

D. L. GREENE

Department of Anthropology, University of Colorado, Boulder

Introduction

A recurring taxonomic problem is the evaluation of the degree and meaning of metrical and morphological variation present within and between fossil species, genera, and families of primates. Consideration of ranges of variation present in accepted modern taxa has in some instances produced considerable simplification and clarification of what seemed to be complex and confusing situations. For example, SIMONS and PILBEAM [1965], using as their standards the ranges of variation observed within and between modern primate genera, were able to argue convincingly that there should be only three genera of hominoids instead of the 26 assigned to the *Dryopithecinae*.

Another area where argument over ranges of variation within and between fossil taxa has led to taxonomic and thereby phylogenetic confusion is the evaluation of the early Pleistocene *Australopithecinae* of South and East Africa. JOHN T. ROBINSON initially argued that the variation within this group of fossils was best explained by the existence of two genera, *Australopithecus* and *Paranthropus*. Recently, ROBINSON [1967], basing much of his newer analysis upon dental metric variation, relegated *Australopithecus* to *Homo (Homo transvaalensis)*, which he then argues gave rise to later, more advanced forms such as *Homo erectus*. He still maintains the distinctiveness and integrity of the genus *Paranthropus*. He feels that the two genera co-existed during the early Pleistocene, with the larger *Paranthropus* occupying a vegetarian, culture-less niche and *H. transvaalensis* occupying an omnivorous niche dependent upon, if not culture, at least tool-use. Ultimately, *Homo* eliminated *Paranthropus* through competition or predation.

MAYR [1951], defining genus as a group of related species that share a major adaptive feature, noted that all of the Pleistocene forms placed in the *Australopithecinae* were probably habitual bipeds, a characteristic held in common with later Pleistocene hominids, and consequently should be placed in the genus *Homo*. Furthermore, he argued that the variation in the fossil record from the early Pleistocene is no greater than that expected within a single primate species and that all forms could be included within a single taxon, *H. transvaalensis.* He noted that sexual dimorphism in some of the living pongids, namely the gorilla and orang, was considerably greater than that found in modern man and that it may be possible to explain the difference between the taxa then designated as *Paranthropus* and *Australopithecus* as being nothing more than a maximum expression of hominoid dimorphism within a single species. Recently, BRACE [1971] has reiterated this dimorphism argument, specifically stating that the differences between *Paranthropus* and *Australopithecus* are no greater than those found between male and female gorilla.

There are a number of other taxonomic positions in the literature [LEAKEY *et al.*, 1964; WOLPOFF, 1971a]; but all can be viewed, for the most part, as extensions and modifications of the reasoning used by either ROBINSON [1967] or MAYR [1951] and BRACE [1971]. That is, the variation observed is similar to that found between the different accepted primate species and genera; or, alternatively, it can be accounted for by reference to degrees of variation that presumably exist within living primate species and, most pertinent, within living hominoid species. Consequently, more systematic evaluations of the variation found within and between living hominoid taxa are needed to help resolve some of the taxonomic controversy.

This paper first describes the nature of dental metric variation found in *Gorilla gorilla gorilla* using data derived from an analysis of specimens from the Cleveland Museum of Natural Science studied by the author. Then it compares this gorilla variation, with a specific emphasis upon sexual dimorphic variation, with that found in the other African pongid, *Pan troglydytes*, and in existing populations of *Homo sapiens*. And finally, using appropriate statistical tests based upon the variation found in living hominoids, it tests the hypothesis that the difference between *Paranthropus* and *H. transvaalensis* in terms of dental metrics is no greater than that to be expected due to sexual dimorphism. This emphasis upon dental metrics is justified by the observation that most extant taxonomic statements in the literature are based upon them.

Gorilla Dental Metric Variation

The gorillas used in this study are 125 lowland specimens (*G. gorilla gorilla*) kept at the Cleveland Museum of Natural Science. According to museum specimen cards, they were collected by professional hunters from two different localities in Cameroon, West Africa. One group is from the vicinity of Ebolowa, approximately one hundred miles south of Yaoundé, the modern capital of Cameroon; and the other comes from Abong Mbang, one hundred miles to the east of the capital. Because of the meager information on the specimen cards, it is impossible to be more specific about the nature of the two geographic samples, though it is probable, since they were collected on contract over a short period of time, that they represent two geographical variants or local populations. The rest of this paper will treat them as distinct local populations.

Age and sex determinations used in this paper follow those recorded by Wingate Todd, the original curator of the collection. Age determinations are based upon dental eruption data and wear, while sex is based upon collector's field observations supplemented by Todd's skeletal evaluation.

A series of standardized measurements [after MOORREES, 1957] were taken from the left side of the jaw for all mandibular and maxillary teeth. Where left teeth were missing, the right isomer was utilized. The measurements taken were:

mesiodistal crown diameter (m-d): the greatest mesiodistal dimension of the tooth crown measured parallel to the occlusal and labial surfaces;

buccolingual crown diameter (b-l): the greatest distance between the labial and lingual surfaces of the tooth crown in a plane perpendicular to that in which the mesiodistal diameter was measured;

buccolingual talonid crown diameter (tal): the greatest distance between the labial and lingual surfaces of the tooth crown within the talonid in a plane perpendicular to that in which the mesiodistal diameter was measured; applicable only to mandibular molars;

buccolingual trigonid crown diameter (tri): the greatest distance between the labial and lingual surfaces of the tooth crown within the trigonid in a plane perpendicular to that in which the mesiodistal diameter was measured; applicable only to mandibular molars;

crown height (h): the distance between the tip of the mesiobuccal cusp and the deepest point of the cemento-enamel junction measured along a line parallel to the long axis of the tooth.

Only unworn teeth were used in obtaining crown heights. The other measurements were taken from teeth with varying degrees of wear, though those worn past the cingular convexity, where maximum crown diameters are usually found, were excluded. Table I summarizes the dental metric statistics.

Since, ideally, taxonomic statements are based upon a large number of characters, the total patterning of character differences within and between taxa is important [GREENE, 1967]. In comparing the gorilla with other hominoid taxa, two kinds of patterning information are useful. One is a description of the metric variation found between gorilla sub-populations and a comparison with that found between sub-populations of other hominoid groups. In such a comparison, males from one gorilla sub-population are compared with males from another gorilla sub-population and females with females. Then, the gorilla male–male variation and female–female variation is compared with other hominoid male–male and female–female variations. This gives an estimate of inter-population dental metric variability found in different hominoid groups. The other kind of patterning information is a description of the metric variation found between the sexes in each gorilla population and a subsequent comparison with that found in other hominoids. This procedure produces comparable estimates of intra-populational sexual dimorphism in dental metrics in different hominoid groups.

Such comparisons can be made in a number of ways. One is to compare the inter- and intra-populational means using critical ratios, if sample sizes exceed 30, and t-tests, if they are less than 30. Usually the results of such tests are indicated by convention as being either significant or very significant. For the first of these two categories, the null hypothesis that the means are the same is rejected at the 0.05 probability level and for the second at the 0.01 probability level. A strict application of critical ratios and t-tests assumes that the frequency distributions of occurrences are normal and in the case of the t-test that sampling variances are equal. Most applied statisticians [SIMPSON et al., 1960] argue that in practice deviations from the assumptions of normalcy and equality of variances have little effect on the sensitivity of these tests.

First, when inter-population variations (using Student's t-test and two-tailed probability, since the appropriate null hypotheses imply no difference between the means being compared) are considered, 4 out of 54 comparisons between Ebolowa and Abong males produced significant differences; while between Ebolowa and Abong females 8 comparisons of 54 were significant (table II). MOORREES [1957], using a critical ratio test,

Table I. Gorilla dental metrics

		Ebolowa						Abong Mbang					
		male			female			male			female		
		n	x̄	±σ	n	x̄	±σ	n	x̄	±σ	n	x̄	±σ
I¹	m-d	13	14.57	1.06	7	13.21	0.78	12	14.59	1.04	14	13.59	0.76
	b-l	13	11.25	1.04	7	10.37	0.79	12	10.93	0.84	14	10.33	0.92
	h	10	13.53	0.49	6	12.76	1.24	11	13.74	1.04	10	13.45	1.36
I²	m-d	12	10.31	0.84	7	9.03	1.00	15	9.99	0.99	15	9.09	1.00
	b-l	13	9.90	1.19	7	9.23	0.52	14	9.86	0.86	15	9.87	1.51
	h	9	12.11	0.75	6	11.01	0.81	11	11.95	0.81	11	11.39	1.21
C¹	m-d	15	21.00	1.91	12	14.14	1.49	14	20.70	3.31	16	15.50	1.29
	b-l	16	14.74	4.42	12	11.32	0.67	14	16.01	1.57	16	11.67	0.88
	h	6	22.86	7.23	9	16.17	2.80	9	29.42	4.17	12	16.85	2.18
P³	m-d	23	11.86	0.98	16	10.94	0.76	24	11.76	0.85	23	11.24	0.62
	b-l	21	15.57	1.12	16	14.58	0.69	24	15.84	1.15	23	14.77	0.69
	h	23	14.99	1.29	16	12.22	1.64	24	14.23	1.41	22	12.57	1.37
P⁴	m-d	22	10.97	1.42	15	10.10	0.53	23	11.15	0.91	22	10.67	0.70
	b-l	22	15.02	1.19	15	13.91	0.53	24	15.22	0.81	22	14.56	0.52
	h	22	11.58	0.95	15	10.23	1.45	24	11.70	1.83	20	10.33	1.08
M¹	m-d	23	15.19	0.91	16	14.25	0.77	23	15.32	0.67	22	14.94	0.54
	b-l	22	15.63	1.40	16	14.79	0.77	23	15.60	0.87	22	14.91	0.75
	h	19	9.65	1.92	15	8.90	1.73	23	9.72	0.88	21	9.92	1.60
M²	m-d	24	16.39	0.99	14	14.56	0.69	25	16.42	1.25	24	15.82	0.91
	b-l	23	16.54	1.46	14	15.19	1.37	25	16.23	1.39	24	15.82	0.64
	h	23	10.19	0.73	13	8.76	0.84	25	10.31	0.67	22	9.64	0.69
M³	m-d	20	14.99	0.90	13	13.92	0.94	19	15.54	1.10	17	14.28	0.73
	b-l	20	15.51	1.08	14	14.19	0.85	19	15.77	1.01	17	14.29	0.77
	h	19	9.81	1.50	14	9.24	2.29	19	9.79	0.95	17	9.22	1.81
I₁	m-d	11	9.06	1.48	6	8.26	0.72	10	8.30	0.47	10	7.71	0.38
	b-l	13	9.32	0.98	5	8.90	0.76	10	9.17	0.45	10	8.72	0.76
	h	9	12.28	1.20	4	11.03	0.79	9	11.88	1.27	8	11.66	1.14
I₂	m-d	15	9.79	0.78	8	8.98	0.87	14	9.36	0.67	14	9.41	1.02
	b-l	16	10.50	1.03	8	9.60	0.85	14	11.03	1.33	15	9.82	0.59
	h	12	14.13	1.41	5	12.79	0.81	10	13.32	1.25	12	12.42	1.42
C₁	m-d	15	15.56	1.35	12	11.08	1.00	15	15.82	1.88	18	11.11	1.32
	b-l	15	18.40	2.01	13	12.55	1.06	15	17.60	2.73	18	12.82	0.88
	h	7	28.42	2.08	8	15.95	2.54	5	27.86	3.80	8	14.55	2.57
P₃	m-d	21	15.87	1.74	16	13.10	1.55	24	16.26	1.30	21	14.14	1.04
	b-l	20	13.76	1.60	16	12.43	1.74	24	13.90	1.08	20	12.22	0.86
	h	19	15.63	1.81	16	12.48	1.77	22	15.81	1.48	20	13.03	1.51

Table I (continued)

		Ebolowa					Abong Mbang					
		male			female			male			female		
		n	x̄	±σ	n	x̄	±σ	n	x̄	±σ	n	x̄	±σ
P₄	m-d	24	12.06	0.81	16	10.83	0.60	24	12.10	0.68	20	11.46	0.49
	b-l	24	13.48	1.20	16	12.28	0.88	24	13.66	1.09	20	12.68	0.84
	h	23	11.66	0.87	15	10.08	1.48	23	11.67	0.74	18	10.31	1.04
M₁	m-d	24	16.11	0.89	15	15.08	0.93	21	16.30	0.72	19	15.50	1.21
	b-l	22	14.02	0.99	15	12.79	0.97	21	13.83	1.07	19	13.22	0.73
	tri	22	13.96	1.11	15	12.83	0.94	21	13.81	1.07	16	13.30	0.64
	tal	21	13.23	0.73	11	12.26	0.88	21	13.41	0.97	17	13.02	0.50
	h	16	9.89	1.23	9	8.43	1.31	21	9.02	0.88	14	8.97	1.67
M₂	m-d	24	17.53	1.08	13	16.05	0.79	23	17.87	1.12	24	17.05	1.27
	b-l	23	15.77	1.14	14	13.06	3.88	23	15.70	1.26	23	14.80	0.63
	tri	23	15.87	1.60	12	14.25	0.82	23	15.69	1.26	23	14.74	0.61
	tal	24	14.83	0.91	12	13.51	0.85	23	15.14	0.96	22	14.46	0.55
	h	19	10.48	1.11	10	8.69	1.09	20	10.16	0.82	21	9.88	1.51
M₃	m-d	20	16.65	1.78	13	15.17	1.57	19	17.45	1.43	17	15.99	0.99
	b-l	19	15.27	0.77	13	13.85	1.01	19	15.42	1.21	17	14.15	0.74
	tri	19	15.18	0.85	13	13.91	0.93	19	15.37	1.36	18	14.11	0.68
	tal	19	13.47	0.99	13	12.04	1.07	18	13.91	1.96	17	12.28	1.26
	h	16	9.98	1.30	11	8.25	1.69	17	9.36	0.84	12	8.10	0.91

compared Aleut dental metric means with those found in seven other distinct modern human populations. His results are not directly comparable to the gorilla comparisons because of his enumerating only critical ratio values of 2.5, which are equivalent to very significant differences. He also did not consider measurements on mandibular incisors and dental heights or mandibular trigonid and talonid dimensions for the other teeth. He found, however, that the male Aleut–male Chinese comparison was the only case that produced very significant differences, specifically, in 11 of 28 comparisons. If the gorilla comparisons are modified by excluding those measurements not considered by MOORREES [1957] and by excluding significant differences, none of the 28 comparisons between Ebolowa and Abong males are very significant; while 9 of the 28 Ebolowa and Abong female comparisons are very significant.

These comparisons indicate that the probability of obtaining very significant differences between means in different gorilla populations falls

Table II. Inter-population variation between two gorilla populations (Ebolowa and Abong Mbang)

		Male–male				Female–female			
		df	t	p		df	t	p	
I^1	m-d	33	0.06		*	23	1.01		+
	b-l	23	0.82			19	0.09		
	h	19	0.54			14	0.94		
I^2	m-d	25	0.85		*	20	0.01		+
	b-l	27	0.06			19	1.68		
	h	18	0.45			15	0.66		
C^1	m-d	27	0.29		+	26	2.48	0.05	+
	b-l	28	0.98		+	26	1.10		+
	h	13	2.09	0.05		19	0.59		
P^3	m-d	45	0.35		+	37	1.30		+
	b-l	43	0.79		+	37	0.81		+
	h	46	1.90	0.05		36	0.71		
P^4	m-d	43	0.41		*	36	2.60	0.01	
	b-l	42	0.65		+	35	3.62	0.01	
	h	44	0.27			33	0.22		
M^1	m-d	44	0.31		*	36	3.14	0.01	+
	b-l	43	0.08		+	36	0.47		+
	h	40	0.15			33	1.78	0.01	
M^2	m-d	47	0.09		*	36	4.36	0.01	+
	b-l	36	0.75		*	36	1.88	0.05	+
	h	37	0.55			33	3.24	0.01	
M^3	m-d	37	1.70	0.05	*	29	1.16		+
	b-l	37	0.75		+	29	0.35		+
	h	36	0.05			29	0.04		
I_1	m-d	19	1.47		*	14	1.89		+
	b-l	21	0.43			13	0.40		
	h	16	0.65			10	0.92		
I_2	m-d	27	1.54		+	20	0.97		+
	b-l	28	1.19			21	0.68		
	h	20	1.34			25	0.52		
C_1	m-d	28	0.42		+	28	0.07		+
	b-l	28	0.88		+	19	0.73		+
	h	10	0.30			14	1.02		
P_3	m-d	33	0.84		*	35	2.37	0.05	+
	b-l	42	0.39		+	34	0.46		+
	h	39	0.34			34	0.98		

Table II (continued)

		Male–male				Female–female		
		df	t	p		df	t	p
P₄	m-d	36	0.18		+	34	3.37	0.01	+
	b-l	36	0.52		*	34	1.32		+
	h	44	0.04			31	0.49		
M₁	m-d	33	0.75		+	32	1.08		+
	b-l	41	0.59		+	32	1.43		+
	tri	41	0.44			29	1.56		
	tal	40	0.68			26	2.84	0.01	
	h	35	2.45	0.05		21	0.78		
M₂	m-d	45	1.02		*	35	2.49	0.01	+
	b-l	44	0.18		+	45	2.05	0.05	+
	tri	44	0.42			43	1.93	0.05	
	tal	45	1.12			32	3.82	0.01	
	h	37	0.17			29	2.16	0.05	
M₃	m-d	37	1.51		+	29	1.70	0.05	+
	b-l	36	0.45		+	28	0.91		+
	tri	36	0.52			29	0.69		
	tal	35	0.85			28	0.53		
	h	31	1.58			21	0.25		

+ = Critical ratio comparison made by MOORREES [1957] between human populations;
* = critical ratio exceeding 2.5 between male Aleut and male Chinese.

within the probability range found in human population comparisons. However, the pattern is reversed between the sexes. In the Aleut–Chinese comparisons, it is the males in one human population that differ from males in another; while in the gorilla population comparisons, it is the females that differ from other females.

Secondly, when intra-population sexual dimorphic variations are considered with the use of Student's t-test and two-tailed probability, 8 of 54 comparisons of Abong Mbang were significant and 29 very significant; while within Ebolowa, 11 of 54 were significant and 34 very significant (table III). MOORREES [1957] provides a critical ratio analysis of the intersex metric differences for his Aleut population using a sample of 28 dental dimensions. Five intersex comparisons were significant; while 11 were very significant. When the gorilla metrics are reduced to the sample used by MOORREES, within Abong Mbang 7 comparisons are significant and 18 very significant; and within Ebolowa, 8 are significant and 20 very signifi-

Table III. Intra-population variation of sexual dimorphism with two gorilla populations

	Ebolowa				Abong Mbang		
	df	t	p		df	t	p
I^1 m-d	19	2.82	0.05	**	24	2.72	0.05
b-l	18	1.86			24	1.65	
h	18	1.64			19	0.53	
I^2 m-d	19	2.85	0.05	**	28	2.38	0.05
b-l	19	1.35			27	0.02	
h	13	2.52	0.05		20	1.21	
C^1 m-d	25	9.83	0.01	**	28	5.61	0.01
b-l	26	2.56	0.05	**	28	9.19	0.01
h	13	2.33	0.05		19	8.49	0.01
P^3 m-d	37	3.06	0.01	**	45	2.38	0.05
b-l	35	3.01	0.01	+	45	3.76	0.01
h	37	5.76	0.01		44	1.41	
P^4 m-d	35	2.21	0.01	+	43	1.80	
b-l	35	3.30	0.01	+	44	3.20	0.01
h	35	3.32	0.01		42	2.88	0.01
M^1m-d	37	3.28	0.01	**	43	2.07	0.05
b-l	36	2.12	0.05	+	43	2.78	0.01
h	42	1.58			42	0.50	
M^2m-d	37	5.93	0.01	+	47	1.90	0.05
b-l	36	2.73	0.01	+	47	1.27	
h	34	5.22	0.01		45	3.29	0.01
M^3m-d	31	3.19	0.01	+	34	3.90	0.01
b-l	32	3.67	0.01	+	34	4.76	0.01
h	31	0.83			34	1.17	
I_1 m-d	15	1.17		*	18	2.95	0.01
b-l	16	0.81			18	1.51	
h	11	0.03			15	0.35	
I_2 m-d	21	2.19	0.05	**	26	0.17	
b-l	22	2.03	0.05		27	3.12	0.01
h	25	1.87			20	1.50	
C_1 m-d	25	9.21	0.01	**	31	8.17	0.01
b-l	26	9.09	0.01	**	31	6.80	0.01
h	13	9.60	0.01		11	6.92	0.01
P_3 m-d	35	4.89	0.01	*	33	5.87	0.01
b-l	34	2.33	0.05	**	42	5.49	0.01
h	33	5.03	0.01		40	5.87	0.01

Table III (continued)

	Ebolowa				Abong Mbang		
	df	t	p		df	t	p
P₄ m-d	38	5.07	0.01	+	42	3.42	0.01
b-l	38	3.34	0.01	*	42	3.23	0.01
h	37	4.02	0.01		39	4.75	0.01
M₁ m-d	37	3.33	0.01	**	38	2.48	0.05
b-l	35	3.63	0.01	*	38	2.05	0.05
tri	35	3.13	0.01		35	1.65	
tal	30	3.20	0.01		36	1.45	
h	23	2.66	0.05		33	0.11	
M₂ m-d	35	4.22	0.01	+	45	2.30	0.05
b-l	35	3.05	0.01	*	45	2.98	0.01
tri	33	3.19	0.01		45	3.17	0.01
tal	34	4.05	0.01		43	2.82	0.01
h	27	4.02	0.01		39	0.73	
M₃ m-d	31	2.38	0.05	+	34	3.42	0.01
b-l	26	4.36	0.01	+	34	3.63	0.01
tri	30	3.86	0.01		35	3.44	0.01
tal	30	3.77	0.01		34	2.84	0.01
h	25	2.88	0.01		35	3.72	0.01

+ = Critical ratio comparison made by MOORREES [1957] between male and female Aleuts; * = critical ratio exceeding 1.96 between male and female Aleuts; ** = critical ratio exceeding 2.5 between male and female Aleuts.

cant. It is clear that the pattern of gorilla intersex differences is not equivalent to that found in the Aleut. Of the Aleut comparisons, 51% are either significant or very significant; while the Abong gorillas show 96% and Ebolowa 100% significant or very significant comparisons. The probability of obtaining a significant sexual dimorphic difference in gorillas is roughly twice that in Aleuts.

Another way of comparably assessing the magnitude of gorilla sexual dimorphism, instead of estimating the probability of obtaining a significant or very significant difference between the sexes in gorillas as opposed to other groups, as done above, is to form a ratio of female mean values to male mean values for the gorilla and comparable groups, as shown in table IV. It is clear that both the gorilla and chimpanzee show much greater dimorphism in the canine dimensions than comparable existing human pop-

Table IV. Comparison of female mean to male mean ratios in five nonhuman and human primate populations

	Ebolowa	Abong Mbang	Chim- panzee[1]	Tristanites[2]	Japanese[2]
I^1 m-d	0.907	0.931	1.000	0.979	0.965
I_1 m-d	0.912	1.006	1.000	0.991	0.963
C^1 m-d	0.673	0.749	0.825	0.976	0.949
C_1 m-d	0.712	0.702	0.809	0.961	0.943
C^1 h	0.707	0.573	0.719		
C_1 h	0.561	0.522	0.771		
P^3 m-d	0.923	0.955	0.989	1.009	0.986
P^3 b-l	0.937	0.932	0.973	0.968	0.979
P_4 m-d	0.898	0.947	0.983	0.989	0.986
P_4 b-l	0.911	0.928	0.975	0.995	0.964
M^1 m-d	0.938	0.975	0.993	0.978	0.980
M^1 b-l	0.946	0.956	0.983	0.970	0.965
M_1 m-d	0.936	0.951	0.979	0.981	0.973
M_1 b-l	0.912	0.956	0.978	0.961	1.001

1 Computed from data presented in SCHUMAN and BRACE [1954].
2 Computed from data presented in MOORREES [1957], and selected because the two groups represent the possible range of sexual dimorphism between human populations.

ulations. Furthermore, within the African pongids the gorilla is substantially more dimorphic in canine dimensions than the chimpanzee. However, the chimpanzee is much closer to *H. sapiens* than to the gorilla in terms of the magnitude of sexual dimorphism as measured by the female/male ratio in incisors, premolars, and molars. Unfortunately, this type of comparative procedure cannot be used with reliability to assess the differences found, in that it is difficult, if not impossible, to assign meaningful standard errors to ratios [SIMPSON *et al.*, 1960]. However, it gives a rough approximation of the distribution of the magnitude of gorilla sexual dimorphism as compared to other forms.

Another procedure for assessing the magnitude of sexual dimorphism that will provide estimates of standard errors is based upon the observed distributions of male-female differences in size. This approach is dependent upon mathematical understanding of the distribution of all possible differences between two sets of variables or the sampling distribution of differences [SPIEGEL, 1961].

It can be shown that, if S1 and S2 are statistics computed from two different populations, the mean of all possible differences S1-S2 (denoted by U_{s1-s2}) and the standard deviation of the difference (denoted by σ_{s1-s2}) are given by:

$$U_{s1\text{-}s2} = U_{s1} \cdot U_{s2}$$

and

$$\sigma_{s1\text{-}s2} = \sqrt{(\sigma_{s1})^2 + (\sigma_{s2})^2}.$$

If S1 and S2 are sample means from the two populations and are denoted by $\bar{x}_1$ and $\bar{x}_2$, then:

$$U_{\bar{x}_1 - \bar{x}_2} = U_{\bar{x}_1} - U_{\bar{x}_2} = U_1 - U_2$$

and

$$\sigma_{\bar{x}_1} - \sigma_{\bar{x}_2} = \sqrt{(\sigma_{\bar{x}_1})^2 + (\sigma_{\bar{x}_2})^2}.$$

These relationships hold for finite populations, if sampling is with replacement; and in this case $(\sigma_{\bar{x}_1})^2$ and $(\sigma_{\bar{x}_2})^2$ can be estimated by

$$\frac{\sigma_1^2}{n_1} \text{ and } \frac{\sigma_2^2}{n_2},$$

or the variance of each finite population divided by its size. If n_1 and n_2 are large (equal to or greater than 30), the sampling distribution of the difference in means is normal, no matter what the nature of the distribution of variates in each population.

Given the above formulations, it then follows that the mean difference between male and female metrics can be calculated by finding the difference between the male and female means:

$$\bar{x}_{\male-\female} = \bar{x}\male - \bar{x}\female$$

and the standard error of this difference by:

$$\text{SE } \bar{x}_{\male-\female} = \sqrt{\frac{SD^2\male}{n\male} + \frac{SD^2\female}{n\female}},$$

where $SD\male$ and $SD\female$ are the calculated standard deviations among males and among females and $n\male$ and $n\female$ are sample sizes for males and females equal to or larger than 30.

Since none of the sample sizes for any tooth, either male or female, from Ebolowa or Abong Mbang is equal to or exceeds 30, means and standard deviations from both populations must be combined in order to calculate a mean difference between males and females and to estimate its standard error. This procedure is done in the following way [PEATMAN, 1963]: if $\bar{x}_t$ is the mean of the combined male sample, $\bar{x}_E$ the mean of Ebolowa males, $\bar{x}_A$ the mean of Abong Mbang males, and n_t, n_E, and n_A the respective sample sizes, then:

$$\bar{x}_t = \frac{n_E \bar{x}_E + n_A \bar{x}_A}{n_E + n_A}.$$

If SD_T, SDE_E, and SD_A are the standard deviations of, respectively, the combined sample, Ebolowa, and Abong Mbang, then:

$$SD_T = \frac{(n_E-1)SD_E{}^2 + (n_A-1)SD_A{}^2 + C_E + C_A}{n_E + n_A-1},$$

where $C_E = n_E(\bar{x}_t-\bar{x}_E)$ and $C_A = n_A(\bar{x}_t-\bar{x}_A)$.

The mean of mean values, $\bar{x}_t$, and the mean of the standard deviations, SD_T, calculated in this way are presumably better single value estimates of true gorilla parameters than the single values from either Ebolowa of Abong Mbang. Combined female values are calculated in the same way.

The mean differences and standard errors of these differences for the total gorilla sample, the Liberian chimpanzee, and three representative human populations, Lapps, Aleuts, and Tristanites, as shown in table V, reinforce the conclusions about sexual dimorphism based upon ratios. Gorilla sexual dimorphism far exceeds that found in the chimpanzee or man in every dimension. Chimpanzee differences in dental metrics between sexes when canine dimensions are excluded are closer in absolute values to those found in modern human populations than they are to gorilla dimensions and in many instances are smaller than those found in modern man.

Evaluation of Australopithecinae Dental Differences

WOLPOFF [1971b] has summarized most of the dental metric data available from Pleistocene hominids. He presents summary statistics, means, and standard errors of the mean for basic dental metrics found in the taxa he designates as *Homo robustus* and *Homo africanus*, which are basically equivalent to ROBINSON's [1967] *Paranthropus* and *H. transvaalensis*. This analysis accepts WOLPOFF's [1971a, b] assignment of specific fossils to one or the other of the taxa, which in most cases represents a concensus evaluation of the positions of most authorities. His *H. robustus* taxon includes specimens from Swartkrans, Olduvai *(Zinjanthropus)*, and Java *(Meganthropus)*, while his *H. africanus* taxon includes specimens from Sterkfontien and Olduvai *(Habilis)*.

Metric differences between the taxa can be systematically evaluated by use of the following series of null hypotheses:

(1) H_{01}: mean differences between *Paranthropus* and *H. transvaalensis* are no greater then those found between male and female gorilla.

(2) H_{02}: mean differences between *Paranthropus* and *H. transvaalensis* are no greater than those found between male and female chimpanzee.

(3) H_{03}: mean differences between *Paranthropus* and *H. transvaalensis* are no greater than those found between male and female *H. sapiens*.

An appropriate test is the following for each metric:

$$t = \frac{(\bar{x}_p - \bar{x}_{Ht}) - (\mu_\male - \mu_\female)}{\sqrt{\dfrac{\sigma_\male^2}{n_p} + \dfrac{\sigma_\female^2}{n_{Ht}}}},$$

where $\bar{x}_p$ = *Paranthropus* mean,

$\bar{x}_{Ht}$ = *H. transvaalensis* mean,

$\mu_\male$ = male mean (H_{01} gorilla, H_{02} chimp, H_{03} sapiens),

$\mu_\female$ = female mean (H_{01} gorilla, H_{02} chimp, H_{03} sapiens),

$\sigma_\male$ = male standard deviation (H_{01} gorilla, H_{02} chimp, H_{03} sapiens),

$\sigma_\female$ = female standard deviation (H_{01} gorilla, H_{02} chimp, H_{03} sapiens),

n_p = size of *Paranthropus* sample,

n_{Ht} = size of *H. transvaalensis* sample.

This test is an extension of the type used when a sample mean is compared to a population mean [SPIEGEL, 1961]. Here the sample mean is the mean difference between *Paranthropus* and *H. transvaalensis* dimensions; and the population mean is, in successive tests, the observed mean difference between male and female gorillas, chimpanzees, and *H. sapiens*. Since each null hypothesis assumes that the sample mean differences could be drawn from each base population, the denominator is the standard deviation of the difference in means from the base population.

As has been discussed, the sampling distribution of differences in means is normal when the sizes of the populations compared are equal to or greater then 30, even if the variates in the populations are not normally distributed. Consequently, metric comparisons using this test are restricted to those dimensions represented by male and female samples equal to or greater than 30 for the gorilla, chimp, and sapiens base populations.

Since in most instances the *Paranthropus* and *H. transvaalensis* samples are small, the tests are evaluated by using the t distribution and appropriate degrees of freedom, defined in this case as $n_p + n_{Ht}$, since none of the parameters are estimated from the samples to be tested but are independently generated from the gorilla, chimpanzee, and human base populations.

And finally, in evaluating the t values, directionality must be considered since it is implied in each null hypothesis. That is, each states that the differences between *Paranthropus* and *H. transvaalensis* are no greater than those found between male and female. Consequently, significance is evaluated by using one-tailed probabilities.

The status of each hypothesis can be expressed in the following summary, based upon the data in table VI:

	Comparisons			Significant rejections			Very significant rejections		
	H_{01}	H_{02}	H_{03}	H_{01}	H_{02}	H_{03}	H_{01}	H_{02}	H_{03}
Maxilla	10	10	10	0	0	0	8	10	9
Mandible	10	8	10	0	0	2	0	5	3
Total	20	18	20	0	0	2	8	15	12

None of the null hypotheses are universally rejected for all metric comparisons. However, the patterns of rejection are interesting. In maxillary comparisons, there are a majority of rejections for all three hypotheses. The degree of dimorphism magnitude exhibited between the maxilla of *Paranthropus* and *H. transvaalensis* far exceeds that found in the modern gorilla, chimp, or man as represented by the Lapps. On the other hand, H_{01} is not rejected in any of the mandibular comparisons; while H_{02} and H_{03} are rejected less often in the mandible than in the maxilla. H_{02} is rejected in 5 of 8 comparisons and H_{03} in 5 of 10.

Clearly, the dimorphism between *Paranthropus* and *H. transvaalensis* as measured by mean differences is greater than expected in either the chimpanzee or modern man in both maxillary and mandibular comparisons, and it exceeds that of the gorilla in the maxillary. The magnitude of the differences between the two taxa cannot be explained by sexual dimorphism of the sort found in either modern man or chimpanzee. It is also improbable that the differences can be explained in terms of something comparable to gorilla sexual dimorphism. Aside from the clear rejection of H_{01} in the maxillary dentition, it should be noted that, while 7 of 10 mandibular comparisons show differences not significantly greater than gorilla dimorphism, 3 of these are significantly or very significantly less. That is, they are smaller differences than those expected in gorilla sexual dimorphism. These t results reject another null hypothesis related to H_{01}, which can be labeled H_{01a} and stated thus: the mean differences between *Paranthropus* and *H. transvaalensis* are not smaller than those found between male and female gorilla. In sum, the overall pattern of the magnitude of sexual dimorphism found between the taxa of the australopithecines is totally unlike that found in gorilla sexual dimorphism.

The dental metric differences observed between *Paranthropus* and *H. transvaalensis* cannot be explained as the operation of sexual dimorph-

Table V. Mean differences (±SE) between males and females for five nonhuman and human primate populations

		Gorilla	Chimpanzee[1]	Lapps[2]	Aleut[2]	Tristanites[2]
I[1]	m-d	1.11±0.27	0.00±0.02	0.03±0.07	0.38±0.07	0.18±0.05
	b-l	0.75±0.27				
I[2]	m-d	1.06±0.27	1.40±0.17	0.14±0.08	0.21±0.07	0.06±0.09
	b-l	0.24±0.31				
C[1]	m-d	5.94±0.55	2.50±0.18	0.27±0.04	0.36±0.04	0.19±0.06
	b-l	3.71±0.64		0.51±0.06	0.32±0.08	0.51±0.10
P[3]	m-d	0.70±0.17	0.09±0.08	0.20±0.04	0.19±0.07	0.06±0.08
	b-l	1.06±0.20	0.28±0.08	0.27±0.06	0.13±0.08	0.17±0.11
P[4]	m-d	0.61±0.21	0.13±0.08	0.13±0.04	0.05±0.09	0.05±0.05
	b-l	0.83±0.18	0.24±0.08	0.25±0.06	0.10±0.15	0.13±0.11
M[1]	m-d	0.61±0.17	0.07±0.07	0.30±0.05	0.32±0.12	0.24±0.06
	b-l	0.76±0.21	0.20±0.07	0.50±0.06	0.13±0.14	0.37±0.08
M[2]	m-d	1.05±0.23	0.25±0.08	0.41±0.06	0.16±0.13	0.25±0.11
	b-l	0.79±0.27	0.32±0.08	0.70±0.06	0.08±0.18	0.36±0.13
M[3]	m-d	1.14±0.23	0.45±0.10	0.18±0.08	0.17±0.21	0.02±0.14
	b-l	1.39±0.22	0.23±0.09	0.52±0.10	0.06±0.27	−0.30±0.29
I$_1$	m-d	0.79±0.29	0.00±0.14	0.14±0.04	0.15±0.07	0.05±0.06
	b-l	0.47±0.25				
I$_2$	m-d	0.23±0.25	0.30±0.15	0.13±0.06	0.19±0.05	0.00±0.06
	b-l	1.00±0.26				
C$_1$	m-d	4.60±0.37	2.10±0.15	0.31±0.04	0.49±0.05	0.28±0.06
	b-l	5.30±0.47		0.60±0.04	0.35±0.09	0.52±0.13
P$_3$	m-d	2.39±0.32		0.13±0.04	0.16±0.07	−0.03±0.08
	b-l	1.52±0.30		0.26±0.04	0.24±0.08	0.12±0.09
P$_4$	m-d	0.80±0.15	0.14±0.10	0.15±0.04	0.15±0.08	0.08±0.06
	b-l	1.08±0.22	0.23±0.11	0.24±0.04	0.26±0.11	0.05±0.09
M$_1$	m-d	0.88±0.22	0.24±0.09	0.31±0.06	0.36±0.14	0.21±0.07
	b-l	0.90±0.21	0.23±0.09	0.35±0.06	0.27±0.13	0.14±0.09
M$_2$	m-d	1.00±0.26	0.19±0.10	0.45±0.06	0.03±0.17	0.26±0.10
	b-l	1.59±0.45	0.16±0.09	0.45±0.06	0.29±0.14	0.22±0.10
M$_3$	m-d	1.41±0.36	0.08±0.11	0.33±0.09	−0.17±0.28	0.38±0.19
	b-l	1.32±0.23	0.36±0.10	0.32±0.08	−0.27±0.20	0.48±0.23

1 Computed from data presented in SCHUMAN and BRACE [1954].
2 Computed from data presented in MOORREES [1957].

Table VI. Evaluation of metric differences between taxa by use of series of null hypotheses[1]

Metrics	Fossil forms			Null hypothesis H_{01}: gorilla as base population[2]						
	$\bar{x}_p$- $\bar{x}_{Ht}$	n_p	n_{Ht}	μ♂- μ♀	n♂	n♀	σ♂	σ♀	t	p
P^3 m-d	1.00	23	15	0.70	47	39	0.92	0.68	1.17	
b-l	1.88	22	15	1.06	45	39	1.13	0.69	2.73	0.01
P^4 m-d	2.10	23	16	0.61	45	37	1.16	0.69	7.07	0.01
b-l	2.75	23	15	0.83	46	37	1.00	0.60	10.57	0.01
M^1 m-d	1.28	22	22	0.61	46	38	0.79	0.72	5.59	0.01
b-l	1.23	22	22	0.76	45	38	1.15	0.75	4.20	0.01
M^2 m-d	1.15	22	19	1.05	49	38	1.12	1.02	3.43	0.01
b-l	0.95	22	19	0.79	48	38	1.42	1.01	1.03	
M^3 m-d	2.09	24	11	1.14	39	30	1.02	0.85	4.63	0.01
b-l	1.96	24	11	1.39	39	31	1.04	0.79	2.77	0.01
P_3 m-d	0.10	22	12	2.39	45	37	1.51	1.37	–4.49	–0.01
b-l	0.46	20	12	1.52	44	36	1.32	1.31	–2.22	–0.05
P_4 m-d	1.01	24	11	0.80	48	36	0.74	0.62	0.87	
b-l	1.21	22	9	1.08	48	36	1.29	0.75	0.35	
M_1 m-d	0.70	32	16	0.88	45	34	0.81	1.10	–0.57	
b-l	0.73	32	15	0.90	43	34	1.02	0.86	–0.59	
M_2 m-d	0.44	23	14	0.99	47	37	1.10	1.21	–1.40	
b-l	0.38	23	15	1.60	46	37	1.19	2.53	–0.54	
M_3 m-d	1.47	20	15	1.41	39	30	1.65	1.31	0.10	
b-l	0.53	20	15	1.32	38	30	1.00	0.87	–2.51	–0.01

1 See text, pp. 94–95.
2 Based upon combining the values from Ebolowa and Abong Mbang to represent more accurate estimates than the single values from either population (see text, p. 93).

ism, at least as it operates in the gorilla, chimpanzee, or modern man, in terms of absolute magnitude or patterning of differences. Consequently, it is probable that the differences observed may be attributable to speciation among the australopithecines. Unfortunately, it is not possible to extend the arguments used in this paper to determine whether the differences observed could be attributed to generic distinctions since *G. gorilla* and *H. sapiens* are monospecific, according to most taxonomists, and data are not available on *Pan pygmaeus*. Consequently, it is impossible to generate

Table VI (continued)

Null hypothesis H_{02}:
chimpanzee as base population[3]

$\mu\male-\mu\female$	n♂	n♀	σ♂	σ♀	t	p
0.09	125	94	0.59	0.58	4.70	0.01
0.28	122	94	0.55	0.64	7.90	0.01
0.13	115	82	0.65	0.49	10.78	0.01
0.24	112	82	0.62	0.53	13.33	0.01
0.07	132	94	0.58	0.47	7.60	0.01
0.20	132	93	0.48	0.48	7.12	0.01
0.25	132	97	0.60	0.58	4.88	0.01
0.32	131	97	0.64	0.57	3.33	0.01
0.45	120	88	0.69	0.70	6.46	0.01
0.23	121	89	0.64	0.64	7.42	0.01
0.14	65	63	0.61	0.47	4.61	0.01
0.23	65	63	0.67	0.58	4.08	0.01
0.24	64	64	0.51	0.49	3.02	0.01
0.23	64	64	0.50	0.54	3.03	0.01
0.19	65	63	0.63	0.55	1.27	
0.16	64	62	0.50	0.52	1.29	
0.08	61	56	0.65	0.61	6.49	0.01
0.36	61	56	0.57	0.53	0.19	

Null hypothesis H_{03}:
Homo sapiens as base population[4]

$\mu\male-\mu\female$	n♂	n♀	σ♂	σ♀	t	p
0.20	221	205	0.45	0.43	3.45	0.01
0.27	203	189	0.57	0.55	6.44	0.01
0.13	237	208	0.46	0.43	9.08	0.01
0.25	208	199	0.58	0.56	10.98	0.01
0.30	256	223	0.64	0.45	5.76	0.01
0.50	231	208	0.61	0.99	3.15	0.01
0.41	267	225	0.65	0.60	3.10	0.01
0.70	258	198	0.64	0.70	0.97	
0.18	196	154	0.70	0.75	5.42	0.01
0.52	187	152	0.82	0.99	5.57	0.01
0.13	226	191	0.45	0.41	−0.13	
0.26	226	197	0.45	0.42	1.91	0.05
0.15	232	191	0.46	0.41	3.61	0.01
0.24	217	203	0.44	0.43	3.73	0.01
0.31	228	192	0.60	0.55	1.69	0.05
0.35	246	207	0.63	0.58	1.58	
0.45	245	203	0.64	0.57	−0.03	
0.45	259	218	0.64	0.59	−0.27	
0.33	172	152	0.92	0.74	3.69	0.01
0.32	168	152	0.78	0.74	0.76	

3 Data from SCHUMAN and BRACE [1954, p. 242].
4 Based upon the Lapp data given in MOORREES [1957, pp. 157–166], chosen because their male and female sample sizes exceed 30 for each comparison.

a sampling distribution of differences between species that belong to single hominoid genera.

Summary

The evaluation of australopithecine dental size differences does not support the argument first made by ERNST MAYR and later revived by C. LORING BRACE that the differences between the presumed taxa are due to nothing more than sexual dimorphism. It tends to support the position long held by JOHN T. ROBINSON that there is at least a specific distinction, if not a generic one, between the taxa.

References

BRACE, C. L.: Sex, inadequacy and australopithecine identity conflicts. Proc. 40th ann. Meet. amer. Ass. of phys. Anthrop. Amer. J. phys. Anthrop. *35:* 274 (1971).

GREENE, D. L.: Genetics, dentition, and taxonomy. Univ. Wyo. Publ. *33:* 93–168 (1967).

LEAKEY, L. S. B.; TOBIAS, P. V. and NAPIER, J.: A new species of the genus *Homo* from Olduvai Gorge. Nature, Lond. *202:* 7–9 (1964).

MAYR, E.: Taxonomic categories in fossil hominids. Cold Spr. Harb. Symp. quant. Biol. *15:* 109–117 (1951).

MOORREES, C.: The Aleut dentition (Harvard University Press, Cambridge 1957).

PEATMAN, J. G.: Introduction to applied statistics (Harper & Row, New York 1963).

ROBINSON, J. T.: Variation and the taxonomy of early hominids; in DOBZHANSKY, HECHT and STEERE Evolutionary biology, vol. 1, pp. 69–100 (Appleton-Century-Crofts, New York 1967).

SCHUMAN, E. L. and BRACE, C. L.: Metric and morphological variations in the dentition of the Liberian chimpanzee. Human Biol. *26:* 239–268 (1954).

SIMONS, E. and PILBEAM, D.: Preliminary revision of the Dryopithecinae. Folia primat. *3:* 81–152 (1965).

SIMPSON, G. G.; ROE, A. and LEWONTIN, R. C.: Quantitative zoology (Harcourt, Brace & Co., New York 1960).

SPIEGEL, M. R.: Theory and problems of statistics (Schaum Publishing Company, New York 1961).

WOLPOFF, M. H.: Is the new composite cranium from Swartkrans a small robust australopithecine? Nature, Lond. *230:* 398–401 (1971a).

WOLPOFF, M. H.: Metric trends in hominid dental evolution (Case Western Reserve University Press, Cleveland 1971b).

Author's address: Dr. DAVID LEE GREENE, Department of Anthropology, University of Colorado, *Boulder, CO 80302* (USA)

Symp. IVth Int. Congr. Primat., vol. 3: Craniofacial Biology of Primates,
pp. 101–127 (Karger, Basel 1973)

Reduction of the Cingulum in Ceboidea

W. G. Kinzey

Department of Anthropology, City College of the City University of New
York

Introduction

The platyrrhine or ceboid primates have long been regarded as representing a stage in primate evolution between the prosimians and the ca-
tarrhine primates. They have frequently been neglected in favor of the
latter because they are not directly related to human phylogeny. Only in
recent decades have they been the object of more intense research, either
because of a unique applicability to a medical problem in man [BULLOCK
et al., 1969; COOPER, 1968; LEVY *et al.*, 1971], or because of an awareness of striking parallelisms between New and Old World primate mor-
phophysiology [e.g., ERIKSON, 1963; KINZEY, 1971]. The extent to which
similarities between Catarrhini and Platyrrhini are due to common ancestry or parallel evolution cannot be determined with our present knowledge of the fossil history of the New World primates. It is hoped, however, that a detailed study of dental characteristics in platyrrhine primates
will pave the way to a clearer understanding of the relationship between
the Ceboidea and the various Eocene prosimians, from one, or perhaps
more, of which the Ceboidea must have arisen.

Very little intensive research has been undertaken on dental variability in platyrrhine primates. Even less consideration has been given to the
phyletic importance of the cingulum in the Order Primates. In his extensive article on primate dentition, REMANE [1960] provided no quantitative data on the presence of this structure; quantitative data are available
only for the gibbons [FRISCH, 1967] and for the upper molars in man
[KORENHOF, 1960].

This paper is an attempt to provide a detailed description and quantification of the cingulum in molar and premolar teeth in living platyrrhine primates. To some degree, particularly in the maxillary lingual cingulum, the living platyrrhines offer an example of a graded series that suggests an evolutionary trend. Obviously, all the living platyrrhines represent endpoints of evolutionary lines that have been separate for varying lengths of time; however, delineation of a pattern of change, such as that of cingulum reduction as represented in the living species, should assist in the proper allocation of fossil platyrrhines.

Throughout the paper reference is made to fields of cingular development. 'The term field implies a region throughout which some agency is at work in a coordinated way, resulting in the establishment of an equilibrium within the area of the field' [HUXLEY and DE BEER, quoted in BUTLER, 1939, p. 2]. The cingulum, or various elements of it, has its maximum expression on a particular tooth. This is most often the anterio-most premolar or the first molar. A field of diminishing intensity of cingulum expression extends mesially and distally. The same homologous structure may have maximum development on different teeth in different species. This variation probably indicates the end result of different evolutionary lines or patterns and assists in tracing relationships among species. The lack of any field in the expression of a trait probably indicates a particular genetic anomaly, unrelated to a general trend. Such is probably the case, for example, in the expression of a protostyle (Carabelli's cusp) on the second molar in *Cebus apella* and in *Lagothrix lagothricha*.

Materials and Methods

Observations were made through a stereo-microscope on approximately 500 skulls and jaws of platyrrhine primates, with samples of 15 to 40 in each of 17 species representing all living genera except *Leontideus*, *Cebuella*, *Callimico*, and *Brachyteles*. Most specimens were from the collection of the American Museum of Natural History; some were from the Field Museum of Natural History. A list of the species and sample sizes may be found in tables I, II, and III. The shape, extent, and position of the cingula and their derivatives are described; and data are presented as percentage presence or absence of various cingular features. Nomenclature of genera and species follows NAPIER and NAPIER [1967]. Heavily worn or broken teeth were excluded. All observations were made on the left side unless teeth were missing, in which case the observation was made on the right since cingular features are expressed more or less similarly on the left and right sides of the jaws. No attempt was made to determine whether any feature was expressed bilaterally or unilaterally.

Tooth Nomenclature

Figures 1 and 2 illustrate and summarize the terminology used to describe the cingula and their derivatives in this paper. Most of the terms are standard, except for one new term (postprotostyle) proposed to identify a specific area of the lingual cingulum; the same terms are used for homologous parts of the cingula on the molars and premolars.

On the upper molars and premolars (fig. 1), the parastyle and the distostyle are the most mesial and distal ectostyles respectively on the buccal cingulum. (The more usual term metastyle is not used for distostyle because it would be descriptively inappropriate on most of the premolars.) The parastyle is mesiobuccal to the paracone on both premolar and molar; the distostyle is distobuccal to the paracone on the premolars and distobuccal to the metacone on the molars. The mesostyle is associated with the groove on the molars between the paracone and the metacone and may lie within the groove or immediately mesial or distal to it. On the premolars the mesostyle is on the midbuccal surface.

Two areas of the upper lingual cingulum are delineated. The protostyle (protostylar area) is the traditional area of the 'Carabelli cusp' mesiolingual to the protocone. I do not use that term for any cusp directly lingual to or distolingual to the protocone. Instead, the terms 'postprotostyle' and 'postprotostylar area' are proposed for the cingular area immediately distolingual to and distal to the protocone, toward the distal end of which the hypocone develops. Thus, the hypocone may be said to develop in the postprotostylar area. When a hypocone is present, there is often a remnant of the cingulum between it and the protocone. The term 'postprotostylar area' is then used for the area of the cingulum between the protocone and the hypocone. Any endostylar cusp which develops between the protocone and hypocone is termed a postprotostyle. Remane [1960] used the term 'interconule' for this

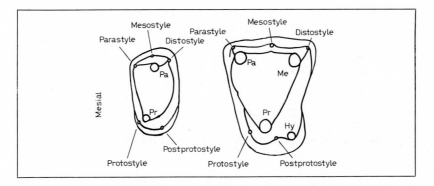

Fig. 1. Hypothetical upper left premolar and molar, illustrating nomenclature of cusps. The protostyle and postprotostyle are derived from the lingual cingulum; the parastyle, mesostyle, and distostyle are derived from the buccal cingulum. Abbreviations: Pr = protocone; Pa = paracone; Me = metacone: Hy = hypocone.

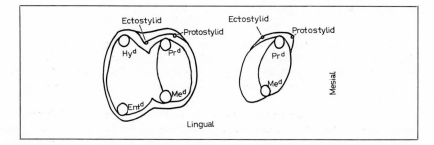

Fig. 2. Hypothetical lower left molar and premolar, illustrating nomenclature of cusps. The ectostylid and protostylid are derived from the buccal cingulum. Abbreviations: Pr^d = protoconid; Me^d = metaconid; Ent^d =entoconid; Hy^d = hypoconid.

cusp; however, it is inconsistent to use conule for a cusp that develops from the cingulum. The same terms are used for homologous positions on the premolars.

On the mandibular premolars and molars (fig. 2), the lingual surface of the enamel does not generally develop a cingulum in primates. Occasionally a small mesiolingual cingulum may develop in the premolars of Eocene Omomyinae, and a lingual cingulum was observed on the lower premolars of a single mandible of *Saguinus mystax*; but these features will not be discussed further.

Stylids develop in two areas of the lower buccal cingulum. Mesiobuccal to the protoconid a protostylid may develop. The buccal cingulum in ceboids forms around the trigonid but does not extend onto the buccal surface of the talonid. A cusp often occurs at the distal end of the cingulum, in or near the groove between the protoconid and the hypoconid. The term ectostylid is used for this cusp. [This is the 'anticonide' of VANDEBROEK, 1969.]

A distinction must be made between primary and secondary cingula. The development of a cusp from a primary cingulum is a major mechanism of cusp formation in mammals. A new or secondary cingulum may then develop around that cusp. Cusps such as the entoconid and hypoconid arise on lower molars from the primary cingulum and hence are termed conids and not stylids [HERSHKOVITZ, 1971, p. 138]. Cusps that in turn develop from the secondary cingulum are termed stylids (styles on upper teeth). This terminology should clarify the situation in the anterior premolars in callithricids. When these are single-cusped teeth, there is a strong primary cingulum around the primary cusp, from which a second cusp, the metacone or metaconid, may develop. In this paper I shall be concerned only with a secondary cingulum, or neocingulum, which develops around the second cusp.

Buccal Cingulum on Lower Molars and Premolars

The mandibular buccal cingulum is the most variable of the dental cingula in platyrrhine primates. The tooth with maximal expression of the field is similarly variable. This variability probably reflects a variation in the function of the cingulum on different teeth in various species. The degree of development of the buccal cingulum in the Ceboidea varies from a completely continuous cingulum around the trigonid (mainly in the Callithricidae) to a complete lack of any cingular derivative (in some Atelinae) (see table I). There is no evidence of continuation of the cingulum distally to lie buccal to the talonid in any living ceboid.

In *Callithrix,* the cingulum is most frequently continuous on M_2 (see table I); but the largest or widest cingulum may be on either M_1 or M_2 (see fig. 4a). Usually, except on M_2, the cingulum is reduced to an ectostylid or ectostylids; and less frequently there is a remnant of the cingulum mesiobuccal to the protoconid. The ectostylid is best developed on P_2, and the protostylid is best developed on M_2.

In *Saguinus,* the cingulum may be continuous in as many as 90 % of jaws; but the tooth with the highest incidence varies among species. The cingulum is often heavily worn on the molars (fig. 4b). The mesiobuccal portion is worn by the P^4 or the metacone of M^1, and the distobuccal portion is worn by the paracone of the upper molar. Thus, the protostylar and ectostylar derivatives of the cingulum are related to different functional requirements. The incidence of derivatives of the buccal cingulum on the premolars is very high, especially on P_2. The mesiobuccal derivative on the anterior premolar is in marked contrast to the condition in *Callithrix.* This difference is probably related to a more pronounced canine-honing mechanism [ZINGESER, 1969, 1971] in *Saguinus.* The upper canine in the latter is larger than that in *Callithrix,* and there is proportionately more attrition on the mesiobuccal surface of P_2 in *Saguinus* than in *Callithrix.* The more developed mesiobuccal ledge on P_2 in *Saguinus* undoubtedly serves to strengthen the tooth against wear from the upper canine. (This feature is also well developed in *Chiropotes.* See figure 3, which illustrates attrition of the cingulum on the anterior premolar due to honing by the maxillary canine.) A high incidence of protostylar ledges in P_3 and P_4 may be the result of field effect.

In *Callicebus* and *Aotus,* the buccal cingulum is not continuous; but derivative ectostylids are often present. In *Aotus* the buccal surface is usually smooth, and protostylids are rarely present. In about one fifth of specimens the first molar has a small ectostylar ledge sloping distobuccally from the protoconid; it is less well developed on the premolars and other molars. In *Callicebus,* a single ectostylid is frequently present on the molars (fig. 4d); and it is better developed, as a rule, on M_2 than on M_1.

Table 1. Buccal cingulum on mandibular molars and premolars. The percentage occurrence (a) of the cingulum as a continuous structure from the mesiobuccal surface of the protoconid to the groove between the protoconid and the hypoconid; (b) of a remnant of the cingulum mesiobuccal to the protoconid (mesiostylid or mesiobuccal cingulum); and (c) of a remnant distobuccal to the protoconid (distostylid or distobuccal cingulum)

Species	N	(a) Continuous						(b) Mesial to protoconid						(c) Distal to protoconid					
		P_2	P_3	P_4	M_1	M_2	M_3	P_2	P_3	P_4	M_1	M_2	M_3	P_2	P_3	P_4	M_1	M_2	M_3
Callithrix argentata	31	3	3	10	24	75	XX	3	10	17	28	75	XX	93	83	79	69	82	XX
Saguinus mystax	32	29	0	0	90	82	XX	100	93	87	90	86	XX	100	93	87	100	97	XX
S. oedipus	17	62	46	25	29	42	XX	100	100	92	21	36	XX	100	100	100	100	90	XX
Callicebus torquatus	40	0	0	0	0	0	0	0	0	0	0	0	0	24	19	8	56	83	10
Aotus trivirgatus	34	0	0	0	0	0	0	5	5	0	0	0	5	5	14	10	18	5	5
Saimiri sciureus	32	0	0	0	41	37	3	43	53	69	97	88	76	53	63	94	100	100	21
Pithecia monachus	29	0	0	0	0	0	0	48	21	14	17	23	38	48	10	0	13	3	0
P. pithecia	29	0	0	0	0	0	0	53	39	22	26	41	63	25	11	0	7	0	0
Chiropotes satanus	27	0	0	0	0	0	0	68	35	46	70	55	37	0	4	8	57	27	16
Cacajao rubicundus	17	0	0	0	0	0	0	31	13	0	6	6	6	75	4	0	0	0	0
Cebus apella	27	0	0	0	0	0	0	46	4	0	0	0	0	50	31	8	4	0	0
C. capucinus	28	0	0	0	0	0	0	22	4	4	0	0	0	30	22	13	0	0	0
C. albifrons	29	0	0	0	0	0	0	26	11	7	3	3	0	33	22	11	0	0	0
Alouatta senicula	28	0	0	0	0	0	0	4	4	4	0	0	0	7	7	18	18	4	0
Lagothrix lagothricha	29	0	0	0	0	0	0	4	0	0	0	0	0	7	0	0	0	0	0
Ateles paniscus	15	0	0	0	0	0	0	0	0	0	0	0	0	0	0	0	7	7	0
A. geoffroyi	23	0	0	0	0	0	0	0	0	0	0	0	0	0	0	0	0	0	0

XX = Tooth absent; N = sample size.

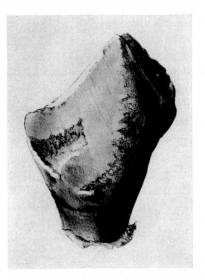

Fig. 3. Buccal view of lower left anterior premolar in *Chiropotes satanus*, illustrating the buccal cingulum partially worn by honing of the maxillary canine tooth.

On M_3 it is only present as a crenulation of the buccal enamel surface. The absence of protostylids in *Callicebus* contrasts with their occasional presence in *Aotus*, and for P_2 this difference reflects the difference in canine utilization in the two genera [KINZEY, 1972]. Canine honing is present in *Aotus*, and the protostylid may be present on the anterior premolars. The canines wear flat in *Callicebus*; honing does not occur, and the protostylid is absent.

In *Saimiri*, the most conspicuous cingular structure is a large stylar shelf, most highly developed on M_1 (fig. 4c). As in the Callithricidae, it may form a continuous shelf around the protoconid. Occasionally, the continuous buccal cingulum may have a small stylid developed upon it, as shown in figure 4c. The field effect is well illustrated in *Saimiri*, with maximum development of the cingulum and its derivatives, both quantitatively and qualitatively, on M_1 and with a reduced incidence and reduced development on each tooth mesial to and distal to M_1. (See table I for the quantitative data.)

In the Pitheciinae *(Cacajao, Pithecia, Chiropotes)*, a continuous cingulum was not observed on any premolar or molar. Only remnants of the cingulum are present, a mesiobuccal shelf and/or a distinct ectostylid (see fig. 4e). The buccal side of the cheek teeth is generally more or less crenulated, especially in *Pithecia;* therefore, only distinct stylids were counted. Typically in the Pitheciinae there is no groove on the buccal side of the mandibular molars between the protoconid and the hypoconid. The lack of ectostylids on the molars in *Cacajao* and the low incidence of ectostylids in *Chiropotes* and *Pithecia* are related to this feature.

In *Alouatta,* there is very little expression of the buccal cingulum on the lower jaw. An ectostylid occurs in or near the groove between protoconid and hypoconid in about one fifth of last premolars and first molars. A mesiobuccal remnant of the cingulum is very rare and occurs only on the premolars. Since P_2 is usually very worn by honing where the protostylid ledge is located, it is possible that its presence on P_2 has been obscured in some cases.

In *Cebus*, derivatives of the buccal cingulum are virtually nonexistent on the molars. On the premolars both mesial and distal ridges are retained to varying degrees. Highest expression, both quantitatively and qualitatively, is on P_2. Unlike the condition of the maxillary cingula in *Cebus*, there is no significant difference among species of *Cebus* in mode or incidence of expression of the mandibular cingulum.

The least expression of the buccal cingulum among the Ceboidea is found in the Atelinae *(Ateles, Lagothrix)*. In some species of *Ateles* (e. g., *A. geoffroyi*), no remnant of the cingulum was observed at all (fig. 4f). The buccal enamel surface of the premolars is usually smooth. On the molars the buccal surface is interrupted by a deep groove between the protoconid and the hypoconid, but only rarely ($7^0/_0$ of M_1 and M_2 in *A. paniscus*) was an ectostylid observed there.

Buccal Cingulum on Upper Molars and Premolars

The maxillary buccal cingulum is also quite variable (see table II). It is rarely retained as a continuous cingulum, but is broken into mesial (parastyle), midbuccal (mesostyle), and distal (distostyle) remnants. The mesial and distal remnants each form a continuous field from premolar series to molar series. The mesostyle, on the other hand, may or may not form separate fields in the premolar and molar series; thus, the molar mesostyle may not always be homologous with the premolar mesostyle.

In *Callithrix* and *Saguinus,* the buccal cingulum is essentially similar.

Table II. Buccal cingulum on maxillary molars and premolars. The percentage occurrence (a) of the cingulum as a continuous structure from parastyle to distostyle; (b) of a remnant of the cingulum mesiobuccal to the paracone (parastyle or mesiobuccal cingulum); (c) of a remnant of the cingulum distobuccal to metacone in molars or distobuccal to paracone in premolars (distostyle or distobuccal cingulum); and (d) of a remnant of the cingulum located midbuccally, between the paracone and the metacone in molars and buccal to the paracone in premolars (mesostyle)

Species	N	(a) Continuous				(b) Parastylar area						(c) Distostylar area						(d) Mesostyle					
		P^2	P^3	P^4	M^1	P^2	P^3	P^4	M^1	M^2	M^3	P^2	P^3	P^4	M^1	M^2	M^3	P^2	P^3	P^4	M^1	M^2	M^3
Callithrix argentata	31	0	0	6	6	100	100	100	100	100	XX	100	100	100	100	94	XX	0	0	7	100	45	XX
Saguinus mystax	32	0	0	3	6	100	100	100	100	97	XX	100	100	100	94	43	XX	21	47	50	100	73	XX
S. oedipus	17	0	0	0	0	100	100	100	100	100	XX	100	100	100	100	100	XX	0	0	7	41	9	XX
Callicebus torquatus	40	0	0	0	0	13	13	8	0	0	0	18	23	23	0	0	0	3	0	0	95	53	0
Aotus trivirgatus	34	9	6	3	0	76	76	73	52	36	9	24	33	30	3	0	4	18	15	12	70	21	0
Saimiri sciureus	32	0	0	0	0	75	59	66	59	47	14	13	19	34	22	13	0	0	0	0	6	0	0
Pithecia monachus	29	0	0	0	0	0	0	0	0	0	0	0	0	0	0	0	0	11	0	0	0	0	0
P. pithecia	29	0	0	0	0	0	0	0	0	0	0	0	0	0	0	0	0	0	0	0	14	0	0
Chiropotes satanus	27	0	0	0	0	0	0	0	4	0	0	0	0	0	0	0	0	0	8	4	4	4	4
Cacajao rubicundus	17	0	0	0	0	0	0	0	0	0	0	0	0	0	0	0	0	56	0	0	0	0	0
Cebus apella	27	0	0	0	0	15	0	4	11	4	0	0	0	0	0	0	0	0	0	0	22	11	0
C. capucinus	28	0	0	0	0	52	17	9	18	4	0	0	0	0	0	0	0	0	0	0	0	0	0
C. albifrons	29	0	0	0	0	0	0	0	3	0	0	0	0	0	0	0	0	4	0	0	0	0	0
Alouatta senicula	28	0	0	0	0	78	100	100	100	100	96	85	100	100	100	100	0	0	0	0	0	0	0
Lagothrix lagothricha	29	0	0	0	0	0	0	0	0	0	0	0	100	100	100	100	0	7	0	0	28	0	0
Ateles paniscus	15	0	0	0	0	0	0	0	20	0	0	0	0	0	0	0	0	0	0	0	20	0	0
A. geoffroyi	23	0	0	0	0	0	0	0	0	0	0	0	0	0	0	0	0	0	0	0	4	0	0

XX = Tooth absent; N = sample size.

It may occasionally form a continuous stylar shelf from the parastyle to the distostyle on P⁴ and M¹ (fig. 5a), often with some small ectostyles rising from the midbuccal surface of the cingulum. More often, however, the continuous cingulum has broken into separate unconnected anterior, midbuccal, and posterior elements (fig. 5c). All three elements normally occur on the molars, but the mesostyles occur with decreasing frequency toward the anterior of the premolar series.

In *Aotus*, the buccal cingulum is a continuous shelf on P² in about one tenth of upper jaws. It occasionally may occur as a continuous structure on P³ and P⁴, but a continuous cingulum was not observed on the molars. Most often the cingulum has broken into two or three remnants (fig. 5e). In *Callicebus*, a continuous buccal cingulum was not observed. A single mesostyle is virtually a constant feature on M¹ and is found in about one half of M² (fig. 5d). A parastyle occurs in about one tenth of the premolars, and a distostyle occurs about twice as often.

In *Saimiri*, there is no evidence of a continuous buccal cingulum. Parastyles and distostyles are common on P² through M². The greatest development may be on either P⁴ or M¹. The mesostyle is rare and was observed only on the first molar and in only two jaws (see fig. 6a).

In the Pitheciinae, the buccal side of the upper premolars and molars is marked by a series of slight crenulations; but there is no evidence of a true cingulum or any ectostylar derivative thereof on the molars (fig. 6e), except that mesostyles occasionally occur in *Chiropotes* and in *Pithecia pithecia*. Typically in the Pitheciinae there is no groove on the buccal side of the enamel between the paracone and the metacone. Only in the smallest species, *P. pithecia*, does a deep groove occasionally occur; and it is in this species that the mesostyle most often occurs, in or near the groove. Its presence there is undoubtedly related to the occurrence of the groove. The anterior premolar has a small mesiobuccal shelf in *P. monachus* (11%) and in *Cacajao* (56%).

Alouatta is remarkable in its development of the buccal cingulum on upper cheek teeth. Elsewhere among the Ceboidea such extensive development is approached only among the marmosets and tamarins. On all P³ through M² (and variably on P² and M³) a wide stylar shelf is developed around the paracone and a separate one around the metacone (fig. 6f). On the premolars the stylar shelves are located mesiobuccal to and distobuccal to the paracone (eocone or amphicone), respectively. There is a well-developed buccal cuspule or ridge between the paracone and the metacone on the molars, which JAMES [1950] and ZINGESER

[1968] refer to as a mesostyle.[1] This cuspule is clearly derived from the ectoloph (eocrista) and therefore should properly be termed an eoconule or mesoloph and not a style (see fig. 6f). It thus serves to divide the buccal surface of the tooth into quite separate mesiobuccal and distobuccal cingula that have not fused into a continuous cingulum. The mesoloph occasionally becomes pinched-off and secondarily attached to the cingulum, giving the appearance of a continuous buccal cingulum with a 'pseudomesostyle'. SMITH [1970] pointed out that the larger species of *Alouatta* have the better developed 'mesostyle', a condition that further attests to the derived stylar position of this cusp.

The differential wear of the two buccal cingula and the mesoloph is quite remarkable. Although the mesoloph is often heavily worn, in a survey of more than 100 skulls of various species of *Alouatta*, the distostylar cingulum was never observed worn, even when the paracone and metacone were extremely so. The parastylar cingulum was never worn except at its mesial edge, and then only on M^1. The mesoloph apparently functions as an additional buccal cusp or crest, particularly important when the paracone and metacone are partially worn. This is an expected functional adaptation to an herbivorous diet. The buccal cingula, however, must have a different function, since they never occlude with the lower teeth. To suggest that they buttress the mesoloph fails to explain their importance on the premolars, where there are no midbuccal stylar elements or mesolophs.

In *Cebus*, the buccal surface of the molars, normally smooth (fig. 6c), may occasionally have a small parastyle (fig. 6d), or a mesostyle, or both. There are marked species differences in their expression within the genus (table II). *C. albifrons* is virtually without cingular derivatives; no mesostyles were observed in *C. capucinus,* but the parastyle is well developed; *C. apella* has both mesostyles and parastyles.

The Atelinae examined have the lowest expression of buccal cingula on the upper molars and premolars.[2] The buccal sides of the premolars and molars, unlike those of the Pitheciinae, are normally smooth (fig. 5 f).

---

1   GREGORY [1922] and SMITH [1970] also refer to this structure as a mesostyle, although SMITH suggests that it is an ectoloph derivative. [Ed.]
2   Dr. KINZEY mentions that *Brachyteles* is not included in his material. ZINGESER [in press] has since the receipt of this article communicated to the author his observations that *Brachyteles arachnoides* is characterized by well-developed buccal cingular elements on the maxillary premolars and molars (especially M^1). [Ed.]

An occasional mesostyle and an occasional parastyle were observed on M¹. In *Lagothrix* (and in *Cebus,* as well), the premolars have small conules mesial and distal to the paracone. These have been treated as plesioconule and stylocone, respectively, since they lie on the ectoloph (eocrista) and not upon the buccal surface of the enamel; therefore, they were not considered.

In summary, a continuous maxillary buccal cingulum is never a constant feature in any living platyrrhine, but it occurs occasionally in the Callithricidae and in *Aotus.* The cingulum is broken into mesial, midbuccal, and distal elements in the premolars and molars. Distal remnants are found in the Callithricidae, Aotinae, and in *Saimiri* and *Alouatta;* mesial remnants are found in these plus *Cebus;* midbuccal elements occur in these plus occasionally in the Pitheciinae and the Atelinae. The distal remnant, when it occurs, is found more or less equally developed on most of the cheek teeth; the parastyle is most strongly developed on P², although a second field seems to have its maximal expression on M¹ in *Cebus;* the mesostyle also has two fields, one centering on P² and one centering on M¹. The general trend from Callithricidae to Cebidae is for reduction of the incidence of cingula and styles, but no single pattern is apparent for the Ceboidea. The function of the maxillary buccal cingulum and its derivatives is less apparent than that of the mandibular cingulum.

Lingual Cingulum on Upper Molars and Premolars

The maxillary lingual cingulum most clearly demonstrates what may be regarded as a series of graded stages among the living platyrrhines (see table III). The Atelinae (*Ateles, Lagothrix*), the most highly evolved, have lost all remnants of the cingulum, at least on the first molar; *Cebus* and *Alouatta* have retained only a postprotostyle; the Pitheciinae have in addition retained a protostyle; *Saimiri* and Aotinae have retained in part a continuous lingual cingulum; only the Callithricidae always have a continuous lingual cingulum on M¹ (see fig. 7). Development of the hypocone parallels the reduction of the cingulum.

The most complete lingual cingulum among living ceboids is found in *Callithrix* (fig. 5a). On M¹ in *C. argentata* it extends as a clearly defined continuous shelf from the midmesial margin of the tooth, around the protocone (to become continuous with the posterior cingulum), all the way to the distostyle on the distobuccal corner of the tooth. It also occurs in a

Table III. Lingual cingulum on maxillary molars and premolars. The percentage occurrence of the cingulum as (a) a continuous structure around the protocone; (b) a remnant mesiolingual to the protocone (protostyle or protostylar cingulum); and (c) a remnant lingual to or distolingual to the protocone (postprotostyle or postprotostylar cingulum); and, (d) the percentage occurrence of a fully developed hypocone

Species	N	(a) Continuous						(b) Protostylar area						(c) Postprotostylar area						(d) Hypocone		
		P^2	P^3	P^4	M^1	M^2	M^3	P^2	P^3	P^4	M^1	M^2	M^3	P^2	P^3	P^4	M^1	M^2	M^3	M^1	M^2	M^3
Callithrix argentata	31	0	14	62	100	28	XX	0	39	97	100	94	XX	0	32	94	100	100	XX	3	0	XX
Saguinus mystax	32	0	0	32	100	53	XX	0	0	32	100	53	XX	0	24	100	100	100	XX	16	0	XX
S. oedipus	17	0	0	36	100	8	XX	0	0	36	100	8	XX	0	40	100	100	100	XX	0	0	XX
Callicebus torquatus	40	0	0	6	91	97	18	0	0	0	100	100	25	53	79	100	100	100	100	100	100	7
Aotus trivirgatus	34	0	0	0	12	4	0	0	0	0	12	4	0	4	31	59	100	88	74	100	100	4
Saimiri sciureus	32	0	34	66	94	63	0	0	34	66	100	63	0	0	94	100	100	100	89	100	100	14
Pithecia monachus	29	0	0	0	4	8	4	0	0	21	100	100	100	0	0	21	100	100	70	100	100	50
P. pithecia	29	0	0	0	0	0	0	0	7	7	100	100	100	0	14	25	100	100	39	100	100	23
Chiropotes satanus	27	0	0	0	0	0	0	0	0	0	83	70	65	0	0	11	91	83	61	100	100	96
Cacajao rubicundus	17	0	0	0	0	0	0	0	0	0	100	87	50	0	0	0	6	0	0	100	100	80
Cebus apella	27	0	0	0	0	0	0	0	0	0	0	11	0	0	19	26	30	26	15	100	103	0
C. capucinus	28	0	0	0	0	0	0	0	0	0	0	0	0	9	26	44	100	52	5	100	100	82
C. albifrons	29	0	0	0	0	0	0	0	0	0	0	0	0	0	14	19	79	71	27	100	100	23
Alouatta seniculus	28	0	0	0	0	0	0	0	0	0	0	0	0	0	0	4	36	30	21	100	100	12
Lagothrix lagothricha	29	0	0	0	0	0	0	0	0	0	0	3	0	0	0	0	0	0	0	100	100	75
Ateles paniscus	15	0	0	0	0	0	0	0	0	0	0	0	0	0	0	0	0	0	22	100	100	45
A. geoffroyi	23	0	0	0	0	0	0	0	0	0	0	0	0	0	0	0	0	0	44	100	100	38

XX = Tooth absent; N = sample size.

well-developed form, but less frequently, on P³, P⁴, and M². There is no (secondary) cingulum on P². Frequently on P³, P⁴, and M², the cingulum divides into separate elements mesial and distal to the protocone, the protostylar and postprotostylar cingula, respectively. No cusps were observed on the cingulum, except that a single first molar had a small hypocone on the cingulum. In *Saguinus*, the lingual cingulum is similarly continuous on M¹ in all skulls (fig. 5b). Compared to that in *Callithrix*, it does not extend as far buccally on the mesial side of the tooth; however, it is a much wider shelf. It is interrupted by a hypocone in 16% of M¹ in *S. mystax*, but a hypocone was not observed in *S. oedipus*. Maximum hypocone development in *S. mystax* is shown in figure 5c. No cingular cusp formation other than the incipient hypocone was observed on *Saguinus*.

In *Callicebus*, the cingulum on M¹ is a continuous serrated shelf extending from near the midmesial margin around the protocone to become continuous with the hypocone (fig. 5d). The most extensive development may be found on either M¹ or M², or they may be equally well developed. When the hypocone is worn, the cingulum may appear to pass lingual to the cusp rather than become continuous with it. Other than the hypocone, no cusp formation was observed in the protostylar area. In *Aotus*, the cingulum is virtually lost in the protostylar area, but is always retained in some form on M¹ in the postprotostylar area. The pattern is quite variable and may consist of a postprotocone either between the protocone and the hypocone (fig. 5e) or directly lingual to the protocone, or it may consist of a ridge sloping mesiolingually from the hypocone to the protocone. Highest expression is always on M¹, and a remnant of the cingulum may be retained in the postprotostylar area of P² through M³. Wear on the cingulum or its derivative in the postprotostylar area is very heavy from the corresponding entoconid.

In *Saimiri*, a continuous lingual cingulum is almost always present on M¹ (fig. 6a) and less often on P³, P⁴, and M². A large cusp may develop on the cingulum either lingual to the protocone or mesiolingual to it (fig. 6b). HILL [1960] described two cusps on the lingual cingulum. I observed two separate cusps, as in figure 6b, on only three first molars and not at all on second molars. REMANE [1960] noted the presence of both pericone (protostyle) and interconule (postprotostyle) in *Saimiri* in addition to the hypocone. A small postprotostyle may be noted in figure 6b.

In the Pitheciinae, both protostylar and postprotostylar areas normally retain cingula or styles. *Cacajao* is peculiar, however, in that there is no remnant of the cingulum in the postprotostylar area (save on two skulls); whereas a protostyle is always retained on M^1 (fig. 6e). The reduced incidence of the postprotostyle in *Cacajao* is related to the generally smoother, steeper-sided lingual aspect of the upper molars. The protostyle, when present, is also smaller in *Cacajao* than in either *Pithecia* or *Chiropotes*.

In *Alouatta*, there is no derivative of the cingulum in the protostylar area. When present, the only remnant of a lingual cingulum is a small postprotostyle, or styles, in the lingual groove between the protocone and the hypocone (fig. 6f). Greatest development is on M^1.

In *Cebus*, there is also no derivative in the protostylar area except on M^2 in three skulls of *C. apella*. The pattern on the molars varies from two well-developed postprotostyles, to a single postprotostyle (fig. 6d), to a wide stylar shelf without any styles extending to the hypocone from a point directly lingual to the protocone (fig. 6c). On the premolars there is usually a single postprotostyle. The accessory cusps do not occur mesiolingual to the protocone [*contra* KUSTALOGLU, 1960] and are therefore probably not properly termed protostyles.

In the observed Atelinae, the premolars and first two molars have smooth lingual surfaces with no trace of cingular derivatives (fig. 5f). On M^3 in *Ateles* a small remnant of the cingulum may be present in the postprotostylar area leading to the hypocone. A single protostyle was observed on one second molar of *Lagothrix;* otherwise, *Lagothrix* is without any remnant of the lingual cingulum.

In summary, the maxillary lingual cingulum in Callithricidae, Aotinae, and *Saimiri* is present usually as a continuous structure around the protocone; and no cusps (except the hypocone) are found on the cingulum, except in *Aotus*. Protostylar remnants are retained in the Pitheciinae, and postprotostylar remnants in Pitheciinae, *Alouatta*, and *Cebus*. The observed Atelinae have the most evolved condition with complete loss of the lingual cingulum. The lingual cingulum exerts a strong field effect with the highest expression almost always on M^1 and with reduced expression mesial and distal to this tooth. When isolated instances of the protostyle occur, as on M^2 in *Lagothrix*, they appear to occur on the largest molar, as in the Hominoidea [KORENHOF, 1960]. Isolated instances of the postprotostyle occur on M^3 in *Ateles*. A hypocone is always present on M^1 and M^2 in the Cebidae and is variably present on M^3. A hypocone does not occur on M^2 in the Callithricidae, but is occasionally present on M^1.

Development of the Hypocone

While the general trend in the Ceboidea, as in the Hominoidea, has been cingulum reduction, there has been a concomitant trend toward development of the hypocone from the postprotostylar area of the maxillary lingual cingulum. (The extreme development of upper buccal cingula in *Alouatta* may be an exception to the general trend of cingulum reduction.) A three-cusped molar pattern is the primitive condition in primate upper molars. In none of the Eocene omomyines or anaptomorphines which might be considered ancestral to the Ceboidea has the hypocone yet developed. The tribosphenic pattern is retained in the second molar of the Callithricidae and usually in the first molar as well. The Callithricidae are the only living members of the Anthropoidea in which the hypocone is not present. This led REMANE [1960] to the belief that the absence of the hypocone in callithricids was probably a reversion rather than a primitive feature. From the postprotostylar area of the cingulum on the first molar of occasional callithricids, however, an incipient hypocone may be found (fig. 5c). The hypocone is developed and retained in the first and second molars of all the Cebidae (fig. 5d, e, f; 6 a–f). The hypocone is exceedingly variable in the third molar, however. In this regard, *Callimico*, usually with a slight hypocone on M^1 and M^2, is aligned with the Cebidae. If the presence or absence of a hypocone on M^2 as a distinguishing character of living cebids and callithricids respectively may be extended to the fossil record, then the Oligocene *Branisella* should be regarded as a cebid. It must be remembered, however, that there was a parallel trend in Oligocene omomyines for development of the hypocone on M^1 and M^2 as seen in *Rooneyia*.

The hypocone may develop either from the lingual cingulum or from a splitting of the protocone. Although the distinction between different origins of the hypocone has no functional significance, it is of tremendous importance in the assessment of phyletic relationships. GREGORY [1922] stated that in *Callicebus* the distolingual cusp on maxillary molars developed not from the cingulum but from a splitting of the protocone, as in *Notharctus*. He called the cusp a pseudohypocone. In actuality, the hypocone in *Callicebus* develops from the lingual cingulum and becomes secondarily attached to the protocone. The evidence for this argument follows.

(1) On dP^4, which is generally more conservative than M^1, the hypocone is no closer to the protocone than it is on M^1.

(2) The pattern of crest attachment to the hypocone varies greatly. Although GREGORY stated that the hypocone crest (entocrista) attaches directly to the protocone, it may also attach to the metaloph between the protocone and the metacone.

(3) In respect to the variability of the location of the entocrista, it is always continuous with the postprotostylar cingulum; it may or may not be continuous with the metaloph.

(4) Intermediate stages in the splitting of the hypocone from the protocone have not been observed in *Callicebus.*

(5) When a hypocone develops on M^3 in *Callicebus* (a rare occurrence), it is clearly upon the cingulum and not connected at all to the protocone.

(6) Enamel endocasts of *Callicebus* molars (which show the pattern of the dentine and a more primitive stage of development than the outer surface of the enamel) show the entocrista connecting the hypocone to the metaloph and not directly to the protocone.

(7) In the closely related *Aotus*, upper molar crown morphology is similar to that in *Callicebus,* but the entocrista either connects the hypocone to the metaloph or has no mesial connection to any other crest or cusp; it does not connect the hypocone to the protocone.

(8) In *Alouatta*, the hypocone is connected to the crest of the protocone (fig. 6f). TARRANT and SWINDLER [1973] recently demonstrated that the last deciduous and the first permanent molars in *Alouatta* each calcify from four separate centers and that the center for the hypocone is the last to develop; it does not develop in relationship to the calcification center of the protocone. A similar study should be undertaken in *Callicebus.*

In all of the ceboid primates, the hypocone develops from the lingual cingulum; thus the term pseudohypocone [as used by GREGORY, 1922, for *Callicebus* and ZINGESER, 1968, for P^4 in *Alouatta*] is inappropriate for any of the New World anthropoid primates.

Evolutionary Trends

A. Evolution of the Mandibular Buccal Cingulum

An anterior buccal cingulum was a primitive condition in the primates, as shown, for example, in the Late Cretaceous–Early Paleocene

Purgatorius [VAN VALEN and SLOAN, 1965]. It is retained and frequently enlarged in the Omomyinae and in the Anaptomorphinae and may extend to both trigonid and talonid, as for example in the Early Eocene *Omomys minutus.* If ceboids ever had a complete buccal cingulum on both the trigonid and talonid of lower molars, there is no evidence of it in living species. In the Miocene *Neosaimiri,* however, the buccal cingulum is more extensive than in the living *Saimiri.* Thus, *Saimiri* represents a reduction of the buccal cingulum from that seen in *Neosaimiri;* and the latter has the expected configuration of the cingulum for an ancestor of *Saimiri.* HERSHKOVITZ [1970a], on the other hand, suggested that the shape of the mandible and the position of the incisors in *Neosaimiri* 'point away from the squirrel monkey'.

The greatest development of the mandibular cingulum in living ceboids is seen in the Callithricidae and in *Saimiri* (fig. 4a, c), in which the cingulum is continuous around the buccal side of the trigonid from the mesiobuccal surface to or slightly beyond the groove between the protoconid and the hypoconid. Thence, reduction takes place by separation of the cingulum into mesial and distal derivatives, the protostylid and the ectostylid. These two elements express themselves differently and have evolved independently in different ways. The retention of a protostylid ledge on P_2 is a function of its importance in canine honing, where it becomes a buttress for the shear of the upper canine against the mesiobuccal surface of the anterior premolar (fig. 3). Thus, it disappears in *Callicebus,* in which canine honing is absent; and it is most strongly developed in those genera in which canine honing is most important. Only in the large platyrrhines such as *Alouatta* and the Atelinae does the functional requirement for this buttress on the P_2 disappear, presumably because of the absolutely larger size of the premolar. A second field for the protostylid is found on the molars, especially in the Pitheciinae.

Except for the general reduction in continuity of the cingulum from Callithricidae to Cebidae, no clear trend in the pattern of reduction is evident. Incidence of the different cingular derivatives probably will be best explained as a response to differing functional requirements.

B. Evolution of the Maxillary Lingual Cingulum

The upper lingual cingulum originated through a fusion of the anterior and posterior cingula. The presence of separate nonfused cingula is

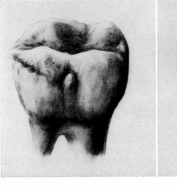

a

b

c

d

e

f

Fig. 4 (Plate I). Bucco-occlusal views of lower left first molars; the mesial side is to the left. (a) *Callithrix argentata*, illustrating continuous buccal cingulum on the trigonid. (b) *Saguinus oedipus*, illustrating continuous buccal cingulum worn through attrition by P4 mesially and by the paracone of M1 distally. (c) *Saimiri sciureus*, illustrating continuous buccal cingulum on the trigonid, with a small ectostylid on the distal end of the cingulum. (d) *Callicebus torquatus*, illustrating an ectostylid in the groove between the protoconid and the hypoconid. (e) *Pithecia monachus*, illustrating both an ectostylid and a protostylid shelf. (f) *Ateles paniscus*, illustrating completely smooth buccal surface without cingular derivatives.

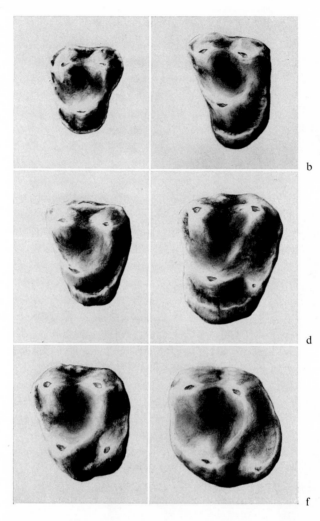

Fig. 5 (Plate II). Occlusal views of upper left first molars showing a graded series of lingual cingulum reduction. The mesial side is to the left, and the buccal side is at the top. (a) *Callithrix argentata*, illustrating completely continuous lingual, distal, and buccal cingula. No lingual cingular cusps are present. (b) *Saguinus oedipus*, illustrating a continuous lingual cingulum, broader distolingually but less extensive mesially than in *Callithrix*, the buccal surface retaining a well-developed parastyle. (c) *S. mystax*, illustrating a continuous lingual cingulum with maximum development of the hypocone observed in the Callithricidae. Well-developed parastyle, mesostyle, and distostyle are present on the buccal cingulum. (d) *Callicebus torquatus*, illustrating a broad continuous lingual cingulum with fully developed hypocone. A mesostyle is present on the buccal surface. (e) *Aotus trivirgatus*, illustrating the

the primitive condition in primates, as seen, for example, in *Purgatorius*. In the earliest Eocene Anaptomorphidae, such as in *Shoshonius cooperi*, the anterior and posterior cingula have not met to form a complete lingual cingulum; but in many Middle and Late Eocene forms the lingual cingulum is complete. The Ceboidea may have evolved from an ancestor that already had developed a complete lingual cingulum, since it is normally present in the Callithricidae, at least on the first molar. *Callithrix*, however, more probably represents a late stage in the formation of the lingual cingulum, since the cingulum is narrow and separate shallow anterior and posterior elements are the usual condition on M². It is more likely that the Ceboidea evolved from a prosimian ancestor that had not yet completed the fusion of anterior and posterior cingula.

A graded series of stages may be seen on the lingual side of upper molars in living ceboids. These are illustrated on first molars in figure 5. Each stage obviously represents the endpoint of a separate evolutionary line; nevertheless, the series illustrates the process by which the ancestors of each probably evolved. In stage I (*Callithrix*, fig. 5a), the anterior and posterior cingula have fused to form a narrow but complete lingual cingulum around the protocone.

Stage II is represented by *Saguinus*, in which the lingual cingulum is complete and has begun to enlarge distal to the protocone (fig. 5b) and an incipient hypocone may occasionally be present (fig. 5c). This stage is also represented by the Bolivian Oligocene *Branisella* [HOFFSTETTER, 1969], although the hypocone is fully developed on both M¹ and M² and *Branisella* may therefore more properly belong in the next stage.

Callicebus represents stage III, in which the hypocone is fully developed on both the first and second molars and the lingual cingulum is still continuous around the protocone and onto the hypocone (fig. 5d).

In stage IV, the hypocone is retained and the cingulum separates into protostylar and postprotostylar elements or remnants. *Aotus*, the Pitheciinae, *Cebus*, and *Alouatta* all are included in this stage; however, they represent separate evolutionary endpoints. *Aotus*, *Cebus*, and *Alouatta* have retained only the postprotostyle (fig. 5e); *Cacajao* has retained only the protostyle (fig. 6e); and *Pithecia* and *Chiropotes* have retained both the protostyle and the postprotostyle. The presence of the protostyle in

---

presence of a postprotostyle between the protocone and the hypocone and of a group of three mesostyles on the buccal surface. (f) *Ateles paniscus*, illustrating complete lack of lingual or buccal cingular derivatives.

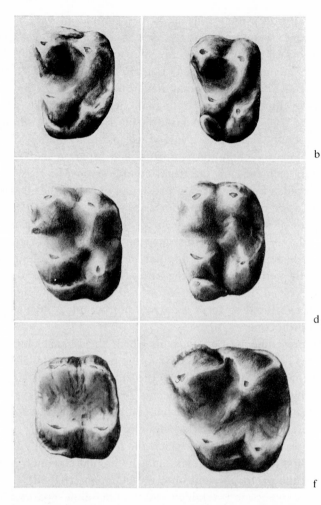

a

b

c

d

e

f

Fig. 6 (Plate III). Occlusal views of upper left first molars; orientation as in figure 5. (a) *Saimiri sciureus,* illustrating an uninterrupted lingual cingulum, continuous with the hypocone. A mesostyle is located on the buccal surface. (b) *S. sciureus,* illustrating a large protostyle and a small postprotostyle on the mesial slope of the hypocone and a small parastyle present on the mesiobuccal side of the paracone. (c) *Cebus albifrons,* illustrating a lingual cingulum in the postprotostylar area leading onto the hypocone. No stylar elements are present on the buccal surface. (d) *C. capucinus,* illustrating a large postprotostyle directly lingual to the protocone and a small parastyle present on the mesiobuccal side of the paracone. (e) *Cacajao rubicundus,* illustrating a protostylar ledge. There are no cingular derivatives on the crenulated buccal surface. (f) *Alouatta senicula,* illustrating a small postprotostyle in the groove between the protocone and the hypocone and the buccal surface contain-

various platyrrhine primates has no phyletic significance as such, as was pointed out by KUSTALOGU [1960]; but, when seen in the context of the trend toward reduction of the lingual cingulum, it assumes a phyletic significance.

In the Miocene *Cebupithecia* [STIRTON and SAVAGE, 1951], the lingual cingulum is well developed on P^4, M^1, and M^2 and is crenulated as in living Pitheciinae. The cingulum is better developed than on any living member of the subfamily, as one would expect of a Miocene representative. HERSHKOVITZ [1970a], however, believes that *Cebupithecia* is unrelated to the living Pitheciinae.

Stage V is the complete loss of all cingular elements except the large hypocone, represented by Atelinae (fig. 5f).

C. Evolutionary Trends in the Platyrrhini

HERSHKOVITZ [1970b], on the basis of a study of endocranial casts of platyrrhine primates, suggested that the 'evolution of cerebral fissural patterns pursues the same pathway in all primates'. A phyletic sequence of living platyrrhines was proposed on the basis of these patterns. The sequence is very similar to that proposed here on the basis of the cingula (see fig. 7). In both HERSHKOVITZ's scheme and this one, *Callithrix* is the most primitive genus, followed by *Saguinus*, followed by the Aotinae, Pitheciinae, *Cebus*, and *Ateles*, in that order. We differ in the placement of *Saimiri* and *Alouatta*. The dental cingulum in *Saimiri* is very primitive, more so than that in *Aotus* and *Callicebus;* whereas the cerebral fissure pattern is more highly evolved. The diet in *Saimiri* is highly insectivorous [FOODEN, 1964] and probably departs relatively little from the dietary pattern seen in *Saguinus*. Consequently, the dental cingula, which evolve in response to functional requirements of mastication, have evolved but little from that seen in the Callithricidae. The cingulum in *Alouatta* is much more highly evolved than its cerebral fissure pattern would suggest. *Alouatta* is the most specialized feeder among the New World primates, eating primarily foliage and unripe figs [HLADIK *et al.,* 1971]; and it is

---

ing mesiobuccal and distobuccal cingula better developed than in any other living platyrrhine primate. A distinctive mesoloph, formed by an eversion of the ectoloph between the paracone and the metacone, appears between the two buccal cingula.

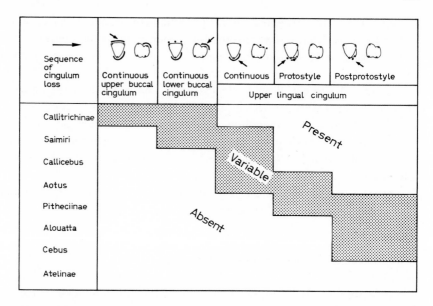

Fig. 7. Occurrence of the cingula on the first molars in Ceboidea. The small arrows point to the cingulum or its derivative considered in each column. The heavy lines represent cingula; the dots signify the breakdown of the cingulum. Each cingulum is considered as a continuous structure; and the lingual cingulum derivatives, protostyle and postprotostyle, are considered separately. A graded series exists from the Callitrichinae, which have continuous lingual cingula on M^1 and may have continuous buccal cingula, to the Atelinae *(Ateles* and *Lagothrix),* in which the cingula and all their derivatives are universally absent from the first molars.

thus not surprising that the cingulum is more highly evolved than that in the (cerebrally) more advanced Pitheciinae.

D. Evolutionary Trends of the Cingulum in the Order Primates

FRISCH [1965] suggested that there was a trend in the reduction of the cingulum in the Hominoidea which followed a fixed pattern: 'The buccal cingulum is lost first, both in the upper and in the lower molars; the lingual cingulum is lost subsequently and apparently much more slowly ...' The same pattern occurs in the Ceboidea (see fig. 7). The details of the reduction differ somewhat from that in the hominoids. In the latter, for example, the lingual cingulum is lost first on the premolars and then on the last molar. This is not the case in the Ceboidea, in which the pattern is a more or less equal loss on all teeth in the premolar-molar series. The cingulum is usually lost first in the protostylar area and subsequently in the postprotostylar area. Unfortunately, the fossil platyrrhine material is not sufficient to support or reject this hypothesis. What little evidence there is, however, tends to support the general trend presented here. Thus, the pattern of cingulum reduction suggested by FRISCH for the Hominoidea is, in its broad outline, more universal among the primates than previously supposed.

Summary and Conclusions

(1) A general trend is observed toward complete loss of the cingula in the Ceboidea as represented in this paper by the graded series of living ceboids from the Callithricidae to the Atelinae. The trend follows a fairly regular pattern when consideration is given to the presence or absence of completely continuous cingula on first molars and to the derivation of separate protostylar and postprotostylar areas on the lingual cingulum.

(2) As in the Hominoidea, the buccal cingula are lost first; and the lingual cingulum (on upper teeth) is lost last.

(3) The buccal cingula tend to be more variable than the lingual cingulum, and a consistent pattern of reduction is not present.

(4) The cingulum is a functional structure and is retained, perhaps enlarged, or lost in response to functional requirements. Only through the field effect is a cingular structure retained with minimal functional requirements. The presence of the maxillary buccal cingulum in *Alouatta* is enigmatic, since its function is not yet understood.

Acknowledgments

Thanks are due to the Department of Mammalogy, American Museum of Natural History, New York, for generous use of its facilities and study of its collections; to Dr. PHILIP HERSHKOVITZ for loan of specimens from the Field Museum of Natural History, Chicago, and for permission to examine the remains of the La Venta Miocene and the Argentine fossil primates on loan to him; also, to Dr. FREDERICK SZALAY for permission to examine the extensive collection of early prosimian fossils and casts in his laboratory; to Dr. ROBERT HOFFSTETTER for a cast of *Branisella boliviana*; to Dr. ALFREDO LANGGUTH for many helpful suggestions and comments; and to Miss ANNA GILLIAM for technical assistance. I am also grateful to Mrs. AUDREY STANLEY for the drawings in figures 3 through 6, and to Mr. BOB HESS for photographic assistance. A grant form the City College Research Foundation provided partial support for this project.

References

BULLOCK, B. C.; LEHNER, N. D. M. and CLARKSON, T. B.: New World monkeys. Prim. Med., vol. 2, pp. 62–74 (Karger, Basel 1969).

BUTLER, P. M.: Studies of the mammalian dentition. II. Differentiation of the post-canine dentition. Proc. zool. Soc., Lond. B *109:* 1–36 (1939).

COOPER, R. W.: Small species of primates in biomedical research. Lab. Anim. Care *18:* 267–279 (1968).

ERIKSON, G. E.: Brachiation in New World monkeys and in anthropoid apes. Symp. zool. Soc., Lond. *10:* 135–164 (1963).

FOODEN, J.: Stomach contents and gastro-intestinal proportions in wild shot Guian-an monkeys. Amer. J. phys. Anthrop. *22:* 227–231 (1964).

FRISCH, J. E.: Trends in the evolution of the hominoid dentition. Bibl. primat., vol. 3 (Karger, Basel 1965).

FRISCH, J. E.: The gibbons of the Malay Peninsula and of Sumatra. A comparative odontological study. Primates *8:* 297–310 (1967).

GREGORY, W. K.: The origin and evolution of the human dentition (Williams & Wilkins, Baltimore 1922).

HERSHKOVITZ, P.: Notes on Tertiary platyrrhine monkeys and description of a new genus from the Late Miocene of Colombia. Folia primat. *12:* 1–37 (1970a).

HERSHKOVITZ, P.: Cerebral fissural patterns in platyrrhine monkeys. Folia primat. *13:* 213–240 (1970b).

HERSHKOVITZ, P.: Basic crown patterns and cusp homologies of mammalian teeth; in DAHLBERG Dental morphology and evolution, pp. 95–150 (University Press, Chicago 1971).

HILL, W. C. O.: Primates, comparative anatomy and taxonomy, vol. 4 (University Press, Edinburgh/New York Interscience Publishers, New York 1960).

HLADIK, C. M.; HLADIK, A.; BOUSSET, J.; VALDEBOUZE, P.; VIROBEN, G. et DELFORT-LAVAL, J.: Le régime alimentaire des primates de l'île de Barro-Colorado (Panama). Folia primat. *16:* 85–122 (1971).

HOFFSTETTER, R.: Un primate de l'Oligocène Inférieur Sud-Américain: *Branisella boliviana* gen. et sp. nov. C.R. Acad. Sci. *269:* 434–437 (1969).

JAMES, W. W.: The jaws and teeth of primates (Pitman, London 1950).

KINZEY, W. G.: Male reproductive system and spermatogenesis; in HAFEZ Comparative reproduction of nonhuman primates, pp. 85–114 (Thomas, Springfield 1971).

KINZEY, W. G.: Canine teeth of the monkey, *Callicebus moloch*: lack of sexual dimorphism. Primates *13:* 365–369 (1972).

KORENHOF, C. A. W.: Morphogenetical aspects of the human upper molar (Uitgeversmaatschappy, Utrecht 1960).

KUSTALOGLU, O. A.: The evolution and significance of the Carabelli cusp; Unpubl. M.A. Thesis, Department of Anthropology, University of Chicago (1960).

LEVY, B. M.; DREIZEN, S.; HAMPTON, J. K., jr.; TAYLOR, A. C., and HAMPTON, S. H.: Primates in dental research. Proc. 2nd Conf. exp. Med. Surg., pp. 859–869 (Karger, Basel 1971).

NAPIER, J. R. and NAPIER, P. H.: A handbook of living primates (Academic Press, London 1967).

REMANE, A.: Zähne und Gebiss; in HOFER, SCHULTZ and STARCK Primatologia, vol. III/2, pp. 637–846 (Karger, Basel 1960).

SMITH, J. D.: The systematic status of the black howler monkey, *Alouatta pigra* Lawrence. J. Mammal. *51:* 358–369 (1970).

STIRTON, R. A. and SAVAGE, D. E.: A new monkey from the La Venta Miocene of Colombia. Compilac. Estud. geol. of. Colomb. *7:* 345–356 (1951).

TARRANT, L. H. and SWINDLER, D. R.: Prenatal dental development in the black howler monkey (*Alouatta caraya*). Proc. 4th Int. Congr. Primat., Portland, Ore., 1972. Amer. J. phys. Anthrop. *38:* 255–260 (1963).

VANDEBROEK, G.: Evolution des Vertébrés de leur origine à l'homme (Masson, Paris 1969).

VAN VALEN, L. and SLOAN, R. E.: The earliest primates. Science *150:* 743–745 (1965).

ZINGESER, M. R.: Characteristics of the masticatory system; in MALINOW Biology of the howler monkey (*Alouatta caraya*). Bibl. primat., vol. 7, pp. 141–150 (Karger, Basel 1968).

ZINGESER, M. R.: Cercopithecoid canine tooth honing mechanisms. Amer. J. phys. Anthrop. *31:* 205–213 (1969).

ZINGESER, M. R.: The prevalence of canine tooth honing in primates. Annu. Meet. Amer. Ass. phys. Anthropol., Boston 1970; cit. in Amer. J. phys. Anthrop. *35:* 300 (1971).

Author's address: Dr. WARREN G. KINZEY, Department of Anthropology, City College of the City University of New York, Convent Avenue and 138th Street, *New York, NY 10031* (USA)

Symp. IVth Int. Congr. Primat., vol. 3: Craniofacial Biology of Primates, pp. 128–147 (Karger, Basel 1973)

Taxonomic and Phylogenetic Uses of the Study of Variability in the Hylobatid Dentition

J. KITAHARA-FRISCH

Sophia University, Tokyo

Introduction

WASHBURN [1954] pointed out almost twenty years ago that while additional fossil finds would certainly always be welcome by the student of human evolution, much more information could be gathered from the already discovered fossil material if only we were better able to read the available evidence. In recent years, it has become more and more widely recognized that an important condition for improving our reading ability is to gain a more accurate knowledge of the variability characteristic of living primates. This statement holds, of course, in particular for the variability of that part of the body most frequently preserved in the fossil record: the dentition.

The inflated taxonomy of fossil hominids since the discovery of *Dryopithecus* by Lartet in 1856, as noted and described by PILBEAM [1966], was no doubt caused to a large extent by the exaggerated value attributed to minor morphological differences in the dentition. For lack of accurate data on dental variability in extant primate species, this inflation could hardly be prevented. Although the present trend in primate paleontology appears to be to set up fewer distinct species, it must not be overlooked that the tendency to underrate the intraspecific character of differences in tooth size and shape is still very much with us, leading still to the gratuitous creation of new species and even genera. The only way to check this tendency is to gather more data on the metrical and morphological variability of living species of primates. Only when adequate data of this sort are available will one be able to follow the most reasonable, though seldom observed, recommendation formulated by SIMONS [1963]:

'Distinctions in dentition in a hominoid specimen, sufficient to warrant desig-
nation of the specimen as the type for a new species, must be at least as great as the
distinctions that occur between species of the closest living relatives of the fossil
form.'

In spite of a number of studies on primate dental variability that have
appeared in the last ten years [FRISCH, 1963; SWINDLER *et al.*, 1963; BIG-
GERSTAFF, 1966; ZINGESER, 1967; HORNBECK and SWINDLER, 1967;
BOOTH, 1971], many potentially available data remain unexplored. The
realization of the need for collecting such data probably accounts for several study projects now under way, as seen in the report on grants for
1970 by the Wenner-Gren Foundation for Anthropological Research.

Among living hominoids, gibbons are particularly suitable for studies
in dental variability. Firstly, the number of extant species is large. Even if
there is still no universal agreement on the exact number of distinct species
[GROVES, 1971], there is certainly more intrageneric variation in *Hylobates* than in any other hominoid primate. Secondly, this large number of
species and races all seem to share a very similar, if not uniform, ecological setting and subsistence pattern. This fact makes the differences between hylobatid species unlike those that distinguish the three ecologically-
separated genera of baboons *(Mandrillus, Theropithecus, Papio),* whose
variability in dentition is understable in terms of diet [JOLLY, 1970,
p. 157]. Those of us who feel with NAPIER [1970, p. xiii] that morphological features wide open to the pressures of ecological selection offer the
least valuable criteria for the systematists at any level lower than that of
Family will thus agree that differences observed between hylobatid species
and races are likely to be of greater taxonomic relevance than is the
case in baboons. Other considerations that make a study of dental variability in the genus *Hylobates* of particular interest stem from the availa-
bility in museums of representative samples for each modern species or
subspecies, with exactly-known places of origin, and from the possibility
of comparing present variability with the trends that are known to have
characterized a phylum whose evolution can be traced back to Oligocene
times.

On the basis of previously published studies [FRISCH, 1963, 1965,
1967a, 1973], this paper will attempt to define dental variability in gibbons at both the intra- and interspecific level and will then proceed to
compare the most variable features in recent forms with the changes that
have taken place during phyletic evolution. By so doing the author hopes
to be able to show by means of examples how a study of dental variability

in primates, particularly in hominoids, may contribute to the clarification of problems in hominid phylogeny.

Intraspecific Variability

Hylobates lar is a particularly convenient species for the study of intraspecific variability, especially since the Asiatic Primate Expedition collected in 1938 in the district of Chiengmai (Thailand) a large sample of a natural population of gibbons. *(H. lar entelloides)* belonging to that species. The data presented here are all based on a study of this population. One metrical and three morphological characteristics are selected for discussion because of the ease in reduplicating these observations and of the possibility of comparing them with data usually reported in the description of fossil specimens.

A. Molar Size

As shown in table III, M^2 has the largest mean dimensions in its group, followed by M^1 and M^3. This sequence, although it is the most common one, is far from being universal. Though M^2 is always the longest and widest (or one of the two longest and widest) molars, M^3 is the shortest upper molar in 65.5% of the cases and is equal to M^1 in 24.5%. In the remaining 10%, M^1 is the shortest molar. In the lower molars, the order of length is much less clear, due to the subequal length of the three molars; $M_1 < M_2 = M_3$, accounting for 25% of all dentitions, is the most

Table I. Condition of the lingual cingulum in the maxillary molars of *Hylobates lar entelloides*

	N	+	−	o
M^1	42	24	13	5
M^2	25	19	4	2
M^3	19	7	10	2

+ = Clearly-marked cingulum; − = weakly-marked cingulum; o = cingulum absent.

Table II. Location of the hypoconulid on the mandibular molars of *Hylobates lar entelloides*

	N	l	c	b	o
M_1	92	19	70	2	1
M_2	81	7	61	8	5
M_3	62	2	21	25	14

l=Lingual location; c=central location; b=buccal location; o=hypoconulid absent.

frequent sequence. As expected, M_3 is the most variable tooth in the group; it is the longest one in 23.2% of the cases and the shortest in 7.2%. This great variability in the relative length of the lower molars constitutes a warning that we are dealing with a section of the dentition that has probably undergone metrical changes in the not-too-distant past, an inference to be checked by analysis of interspecific differences and by study of the fossil record.

B. Development of the Lingual Cingulum on the Upper Molars

As shown in table I, the degree of expression of the lingual cingulum may vary considerably, not only between distinct species or between varieties of the same species, but even within one population. In the table, the cingulum is indicated as weakly marked when only imperfectly separated from the protocone, either because the separating furrow is too thin or because some sections of it have been obliterated.

C. Anterior Lower Premolar

As in all nonhuman hominoid genera, the anterior lower premolar of the gibbon is of sectorial shape; the crown is enlarged and compressed laterally to form a mesial oblique cutting edge that, in occlusion, shears against the upper canine. The variability observed in *H. lar entelloides* concerns chiefly the degree of lateral compression of the crown. Besides

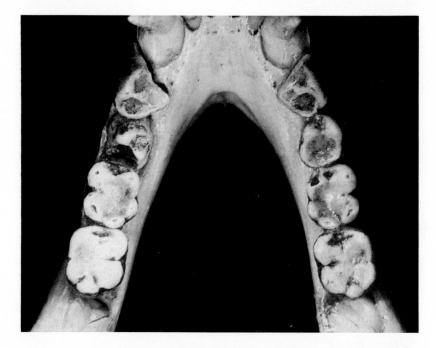

Fig. 1. Lower dentition of *Hylobates lar entelloides,* showing triangular type of anterior premolar and lingually-situated hypoconulid on M_1.

the common or elongated type of crown, another was recorded that is best defined as triangular. In the latter, a strong development of the disto-lingual portion causes the crown to assume a triangular occlusal outline (fig. 1). This variation, however, does not seem to modify the sectorial character of the tooth. Moreover, no lingual cusp was ever found to occur on the expanded distal part; but the latter is often supported by a third root.

Because of the number of forms intermediate between the compressed and the triangular type, the most useful objective criterion appears to be the presence of the third root just mentioned. A triangular P_3 is then found to occur in 26 out of 56 males and in 12 out of 40 females. This high rate of variability is the more remarkable in that it affects a region of the dentition assumed to be quite stable. In view of the almost universal association of P_3 with the canines within a single functional complex, it is also noteworthy that no connection appears to exist in gibbons between the morphological variation in P_3 and the size of the canine teeth.

D. Location of the Hypoconulid

Table II shows the location of the hypoconulid in the lower molars of the *H. lars entelloides* to be much more variable than would be expected from the usual description of the ape molar occlusal pattern as consisting of three buccal and two lingual cusps [e.g., GREGORY and HELLMAN, 1926]. The location of the hypoconulid is said to be lingual, central, or buccal with reference to the mesiodistal length axis of the crown. Especially at variance with the commonly-held opinion of the typical appearance of a lower molar is the small proportion of molars with a clearly buccally-situated hypoconulid. A not inconsiderable number of first lower molars even show a lingually-deflected distal cusp (fig. 1). A trend for the hypoconulid to shift buccally progressively from the first to the last molar is clearly seen. The frequent absence of the hypoconulid on the third molar is also of interest. In many more cases, the hypoconulid is considerably reduced. Since the position and size of the hypoconulid constitute one of the principal elements making up the groove pattern, this pattern appears to be much more variable in the gibbon than it is reported to be in other hominoids. Certainly, the marked variability in the position and size of that cusp invites prudence in interpreting the phylogenetic significance of its position and of the *Dryopithecus* pattern of which it is an element.

Interspecific Variability

A. Molar Size

Table III gives the molar lengths in the listed species and subspecies of *Hylobates* as well as in *Symphalangus*. For completeness, measurements have been segregated by sex (except for the small sample of *H. lar albimanus*). In no case, however, was a statistically significant difference observed between males and females. This situation is in marked contrast with the dentition of many other primates, such as *Alouatta,* where sexually-related dimorphism for size is found in almost all teeth [ZINGESER, 1967].

Table III. Mesiodistal length of molar teeth in various hylobatid species

Species	Tooth	Sex	N	Range	Mean	SD	Tooth	Sex	N	Range	Mean	SD
Hylobates	M¹	♂♂	15	5.3–6.0	5.73	0.27	M₁	♂♂	12	5.6–6.4	6.10	0.22
agilis		♀♀	11	5.2–6.0	5.63	0.25		♂♀	12	5.7–6.4	6.02	0.22
		♂♀	26	5.2–6.0	5.68	0.26		♂♀	24	5.6–6.4	6.06	0.22
	M²	♂♂	15	5.8–6.6	6.19	0.22	M₂	♂♂	13	5.8–6.8	6.38	0.22
		♀♀	11	5.5–6.4	5.93	0.29		♀♀	12	5.7–5.6	6.22	0.28
		♂♀	26	5.5–6.6	6.08	0.28		♂♀	25	5.7–6.8	6.30	0.26
	M³	♂♂	15	4.2–5.6	5.10	0.39	M₃	♂♂	13	5.5–6.4	6.02	0.29
		♀♀	8	4.5–5.5	5.14	0.31		♀♀	10	5.4–6.5	5.83	0.36
		♂♀	23	4.2–5.6	5.11	0.35		♂♀	23	5.4–6.5	5.94	0.33
H. concolor	M¹	♂♂	12	5.8–6.8	6.16	0.26	M₁	♂♂	12	6.2–7.3	6.57	0.34
		♀♀	13	5.8–6.8	6.14	0.32		♀♀	13	6.2–7.3	6.64	0.32
		♂♀	25	5.8–6.8	6.15	0.29		♂♀	25	6.2–7.3	6.60	0.32
	M²	♂♂	10	5.9–7.1	6.27	0.36	M₂	♂♂	10	5.8–7.7	6.62	0.51
		♀♀	13	6.0–7.3	6.49	0.41		♀♀	13	6.3–7.3	6.87	0.34
		♂♀	23	5.9–7.3	6.40	0.40		♂♀	23	5.8–7.7	6.76	0.43
	M³	♂♂	7	4.3–6.7	5.60	0.81	M₃	♂♂	7	6.0–7.0	6.49	0.36
		♀♀	12	4.9–6.5	5.67	0.46		♀♀	12	5.8–7.6	6.57	0.47
		♂♀	19	4.3–6.7	5.64	0.59		♂♀	19	5.8–7.6	6.54	0.42
H. hoolock	M¹	♂♂	20	6.1–7.1	6.63	0.28	M₁	♂♂	19	6.3–7.5	6.88	0.29
		♀♀	20	5.8–7.0	6.41	0.32		♀♀	21	6.1–7.2	6.74	0.33
		♂♀	40	5.8–7.1	6.52	0.32		♂♀	40	6.1–7.5	6.81	0.32
	M²	♂♂	18	6.5–7.9	7.21	0.37	M₂	♂♂	17	7.2–8.3	7.69	0.29
		♀♀	18	6.5–7.7	7.14	0.34		♀♀	19	6.8–8.1	7.51	0.37
		♂♀	36	6.5–7.9	7.17	0.35		♂♀	36	6.8–8.3	7.59	0.35
	M³	♂♂	12	5.7–7.3	6.62	0.53	M₃	♂♂	11	6.8–8.0	7.15	0.34
		♀♀	14	5.9–7.1	6.59	0.40		♀♀	14	6.8–8.0	7.34	0.33
		♂♀	26	5.7–7.3	6.61	0.46		♂♀	25	6.8–8.0	7.26	0.34
H. moloch	M¹	♂♂	20	5.4–6.5	5.93	0.33	M₁	♂♂	19	5.9–7.0	6.32	0.29
(Borneo)		♀♀	18	5.2–5.5	5.93	0.37		♀♀	17	5.8–6.7	6.35	0.22
		♂♀	38	5.2–6.5	5.93	0.34		♂♀	36	5.8–7.0	6.33	0.26
	M²	♂♂	17	5.7–7.2	6.38	0.37	M₂	♂♂	16	5.9–7.3	6.65	0.42
		♀♀	17	5.6–7.0	6.20	0.34		♀♀	16	5.4–7.0	6.47	0.44
		♂♀	34	5.6–7.2	6.29	0.36		♂♀	32	5.4–7.3	6.56	0.43
	M³	♂♂	14	4.4–6.3	5.46	0.53	M₃	♂♂	13	5.4–7.0	6.38	0.56
		♀♀	16	4.6–5.9	5.49	0.42		♀♀	15	5.6–7.1	6.29	0.45
		♂♀	30	4.4–6.3	5.47	0.47		♂♀	28	5.4–7.1	6.33	0.50
H. moloch	M¹	♂♂	10	5.6–6.5	5.83	0.28	M₁	♂♂	9	6.0–6.9	6.20	0.43
(Java)		♀♀	10	5.2–6.3	5.87	0.29		♀♀	10	5.7–6.8	6.24	0.33
		♂♀	20	5.2–6.5	5.85	0.28		♂♀	19	5.7–6.9	6.22	0.38

Table III (continued)

Species	Tooth	Sex	N	Range	Mean	SD	Tooth	Sex	N	Range	Mean	SD
	M²	♂♂	10	5.5–7.0	6.26	0.39	M₂	♂♂	10	6.0–7.5	6.70	0.43
		♀♀	10	5.9–6.8	6.51	0.26		♀♀	10	6.2–7.2	6.72	0.37
		♂♀	20	5.5–7.0	6.38	0.35		♂♀	20	6.0–7.5	6.71	0.39
	M³	♂♂	9	5.0–6.2	5.71	0.41	M₃	♂♂	9	6.0–7.1	6.40	0.34
		♀♀	6	5.0–6.5	5.43	0.56		♀♀	8	5.8–7.7	6.51	0.67
		♂♀	15	5.0–6.5	5.60	0.48		♂♀	17	5.8–7.7	6.45	0.51
Symphalangus	M¹	♂♂	14	6.9–8.3	7.51	0.41	M₁	♂♂	10	6.9–8.1	7.53	0.37
s. syndac-		♀♀	14	6.8–7.9	7.46	0.38		♂♂	12	6.4–8.2	7.32	0.60
tylus		♂♀	28	6.8–8.3	7.49	0.39		♂♀	22	6.4–8.2	7.41	0.51
	M²	♂♂	14	7.8–8.8	8.31	0.26	M₂	♂♂	11	7.6–9.0	8.31	0.43
		♀♀	14	7.5–9.0	8.05	0.47		♀♀	12	6.9–9.7	8.20	0.80
		♂♀	28	7.5–9.0	8.18	0.39		♂♀	23	6.9–9.7	8.25	0.64
	M³	♂♂	12	6.7–8.8	7.67	0.67	M₃	♂♂	10	7.2–9.1	8.00	0.54
		♀♀	14	5.9–8.2	7.19	0.59		♀♀	11	6.9–8.5	7.73	0.46
		♂♀	26	5.9–8.8	7.42	0.66		♂♀	21	6.9–9.1	7.86	0.50
H. lar	M¹	♂♂	6				M₁	♂♂	6			
albimanus		♀♀	2					♀♀	2			
		♂♀	8	5.0–6.1	5.42	0.38		♂♀	8	5.3–6.3	5.60	0.41
	M²	♂♂	6				M₂	♂♂	6			
		♀♀	2					♀♀	2			
		♂♀	8	5.4–6.4	5.80	0.35		♂♀	8	5.5–6.2	5.89	0.27
	M³	♂♂	6				M₃	♂♂	6			
		♀♀	2					♀♀	2			
		♂♀	8	4.0–4.9	4.44	0.24		♂♀	8	4.7–5.4	5.15	0.28
H. lar	M¹	♂♂	43	5.1–6.6	5.75	0.33	M₁	♂♂	39	5.7–6.8	6.19	0.29
entelloides		♀♀	44	5.1–5.9	5.52	0.23		♀♀	41	5.4–6.6	6.00	0.29
		♂♀	87	5.1–6.6	5.64	0.31		♂♀	80	5.4–6.8	6.09	0.30
	M²	♂♂	38	5.2–6.9	6.24	0.36	M₂	♂♂	37	5.8–7.3	6.48	0.35
		♀♀	40	5.4–6.7	6.02	0.32		♀♀	36	5.7–7.2	6.28	0.35
		♂♀	78	5.2–6.9	6.19	0.35		♂♀	73	5.7–7.3	6.38	0.36
	M³	♂♂	31	4.8–6.2	5.36	0.37	M₃	♂♂	30	5.8–7.4	6.44	0.35
		♀♀	30	4.6–5.7	5.13	0.35		♀♀	26	5.5–6.9	6.27	0.40
		♂♀	61	4.6–6.2	5.25	0.38		♂♀	56	5.5–7.4	6.36	0.38
H. lar	M¹	♂♂	9	5.7–6.4	5.90	0.23	M₁	♂♂	9	5.6–6.8	6.29	0.36
pileatus		♀♀	6	5.4–5.8	5.62	0.16		♀♀	6	5.3–6.2	5.90	0.32
		♂♀	15	5.4–6.4	5.79	0.25		♂♀	15	5.3–6.8	6.13	0.39
	M²	♂♂	9	5.6–7.1	6.19	0.42	M₂	♂♂	9	6.0–7.2	6.44	0.40
		♀♀	6	5.5–6.2	5.85	0.30		♀♀	6	5.2–6.5	6.02	0.49
		♂♀	15	5.5–7.1	6.05	0.40		♂♀	15	5.2–7.2	6.27	0.47

Table III (continued)

Species	Tooth	Sex	N	Range	Mean	SD	Tooth	Sex	N	Range	Mean	SD
	M^3	♂♂	9	5.0–6.5	5.53	0.45	M_3	♂♂	9	5.6–6.8	6.13	0.44
		♀♀	4	4.8–5.5	5.22	0.31		♀♀	4	5.0–6.4	5.87	0.61
		♂♀	13	4.8–6.5	5.44	0.43		♂♀	13	5.0–6.8	6.05	0.49

Table III shows that, while within the genus *Hylobates* molar measurements are adequate to distinguish the species or groups with the smallest molars from those with the largest, much overlap occurs between groups having a contiguous geographic distribution (e.g., *H. agilis* and *H. lar albimanus* in Sumatra, *H. agilis* and *H. lar entelloides* in Malaya, *H. lar pileatus* and *H. concolor* in Indochina). Statistical treatment of the data confirms these observations. Student's t-test for the difference between paired samples shows a meaningful difference ($p < 0.05$) to exist only between one pair of neighboring groups, *H. hoolock* and *H. lar entelloides*. Particularly noteworthy, however, is the gradual increase in size observed in several molar teeth from *H. agilis* to *H. lar,* to *H. concolor,* to *H. hoolock*, i.e., as one moves further away from the equator. In spite of a few exceptions,[1] the trend is unmistakable and particularly evident in the third lower molar.

As a consequence of the extensive overlap in molar dimensions, metrical data alone would not allow determination of whether a given tooth found in Sumatra belonged to *H. lar* or to *H. agilis*. An exception should be made, however, for the third lower molar, where no or very little overlap is seen. Metrical reduction of M_3 appears thus to be a species-specific feature. Moreover, since M_3 is relatively easy to identify when found detached from the rest of the dentition, the amount of reduction seen in this tooth may prove to be a useful criterion for the paleontologist. Furthermore, the differential reduction of M_3 according to species confirms the inference drawn from the study of intraspecific variability in the size of the lower molars that metrical changes in this area of the dentition have taken place in the not-too-distant past and may in fact still be operative.

1 More specifically, among the three hylobatid forms that currently inhabit Sumatra, there is nearly complete overlap in molar size between the two forms considered to belong to the same genus, *H. agilis* and *H. lar*. Absence of overlap occurs in Sumatra only between distinct genera, *Hylobates* and *Symphalangus*.

B. Development of the Lingual Cingulum on the Upper Molars

As discussed extensively in a previous publication [FRISCH, 1965], the degree and the mode of reduction of the lingual cingulum on the upper molar dentition differ markedly in the various species of *Hylobates* and may even serve at times to distinguish populations best regarded as belonging to the same species. It is possible to rank the taxa of the genus *Hylobates* according to the degree of this reduction. The resulting sequence is *H. concolor*, almost no reduction; *H. moloch* (Java) and *H. lar pileatus*, little reduction; *H. lar* (other races) and *H. moloch* (Borneo), moderate reduction; and *H. agilis* and *H. hoolock*, extreme reduction.

When this ranking is examined, some interesting characteristics of the cingulum reduction appear. (1) Presence or absence of the cingulum is not related to tooth size; of the two species with the largest molar teeth, *H. hoolock* and *H. concolor*, the former shows the maximum amount of cingulum reduction, the other almost no reduction at all. (2) Significant differences are observed between groupings of gibbons usually considered to belong to the same species; e.g., *H. lar pileatus* shows a very different picture from the other varieties of *H. lar*. Nearly all of the more than twenty specimens of *pileatus* examined, though collected in widely-separated localities, showed a well-developed cingulum on M^1 and M^2; in none was the cingulum entirely absent. In the other races of *H. lar*, the cingulum was clearly marked usually in only one-half of these teeth and was occasionally absent on all three upper molars. Another example is the case of the Javanese gibbons, which differ from the Bornean gibbons in this respect even more sharply than do the races of *H. lar* among themselves, although both Javanese and Bornean populations are commonly classified in the same species *(H. moloch)*.

C. Anterior Lower Premolar

Without showing the aberrant morphology observed in *H. lar entelloides* (although the rare triangular type of premolar similar to that seen in the Chiengmai population was found in a few specimens of *H. lar pileatus*), the anterior lower premolar nevertheless presents a considerable amount of variability in other groups. A more frequent manifestation of the variability found in this tooth, however, is the occasional presence in some species of a cuspule on the crest descending lingually from the major cusp to the robust lingual cingulum. This condition is observed in a number of species, but is too ill-defined to allow calculating meaningful frequencies.

D. Location of the Hypoconulid

The observations summarized in table IV confirm some of the conclusions drawn from the comparative study of the development of the cingulum. Not only is great variation seen in the most frequent position of the posterior cusp for each species of *Hylobates*, but varieties commonly considered to form one species may differ considerably in this respect.

Within *H. lar,* the Chiengmai population *(H. lar entelloides)* is most obviously unique and characterized by the large number of M_1 and M_2 with lingually-shifted hypoconulids. *H. lar pileatus* appears to be the group most resembling the Chiengmai population; *H. lar albimanus* is the

Table IV. Location of the hypoconulid on the lower molars of various hylobatid species[1]

	M_1				M_2				M_3			
	l	c	b	o	l	c	b	o	l	c	b	o
Hylobates agilis	–	7	22	1	–	8	25	–	1	10	7	11
H. concolor	–	11	14	–	–	10	13	1	–	7	11	4
H. hoolock	–	26	22	–	–	27	17	–	–	18	6	10
H. lar albimanus	–	2	8	–	–	6	4	–	–	2	1	6
H. lar lar	–	17	9	–	–	17	6	–	–	10	8	4
H. lar entelloides	18	70	2	1	7	61	8	5	2	21	25	14
H. lar pileatus	3	20	1	1	–	21	1	1	1	9	8	2
H. moloch (Borneo)	–	8	51	1	–	20	32	1	–	7	22	18
H. moloch (Java)	–	14	12	–	–	12	12	–	–	7	9	1
Brachytanites (H. Klossi)[2]	–	3	11	–	–	4	10	–	–	1	2	9
Symphalangus s. syndactylus	–	27	22	–	–	19	25	–	1	16	17	–

1 Symbols as in table II.
2 See footnote 2 in text [Ed.].

most different. It is tempting to see in this contrast a reflection of the geographical distance separating each of these two races from the Chiengmai population. *H. lar pileatus* inhabits the forests of southeastern Thailand, less than 500 km from Chiengmai; *H. lar albimanus*, on the other hand, is found only in the northwest corner of the island of Sumatra, about 1,500 km away, and has presumably been separated from the other populations of *H. lar* by the rising sea level since at least the end of Pleistocene times.

The Javanese and Bornean gibbons are as distinct from each other in the position of the hypoconulid as they were found to be in the degree of expression of the cingulum. It should be noted, however, that these differences are not nearly as marked as those existing among the varieties commonly included in *H. lar*. Thus, since greater discrepancies are found among what are probably best regarded as races of one species, the differences distinguishing Bornean from Javanese gibbons cannot be given specific value, at least not in isolation from other differences.

E. Conclusions

Interspecific dental variability in *Hylobates*, whether it be examined in its metrical or morphological aspects, thus presents a reasonably consistent picture of a continuum where adjacent populations are hardly distinguishable but geographically distant populations are much more easily identifiable. In several instances, distinctions between neighboring varieties are more clearly marked in the morphology than in the measurements (two examples are the Bornean and Javanese varieties of *H. moloch* and *H. agilis* and *H. lar albimanus* on Sumatra); this fact should be of interest to paleontologists. Even then, however, the differences observed would not suffice to identify with reasonable certainty the origin of a specimen, were this not otherwise known. A consequence of this continuum type of variability is that, no matter what taxonomic arrangement is preferred when classifying living gibbons, as much or more morphological and perhaps metrical variability may be found within each species as between distinct species. When one reflects that the situation now existing in gibbons was probably typical of hominoids in the geological past, it seems natural that, as our knowledge of extinct primate species becomes less fragmentary, the paleontologist is facing the same sort of problems already long-experienced by the neozoologist.

Table V. Lower molar length in fossil hylobatids and in *Symphalangus*

	M₁	M₂	M₃
Limnopithecus legetet	5.5–6.0	5.7–7.1	6.4–7.3
L. macinnesi	6.1–7.5	7.2–8.2	8.2–8.9
Pliopithecus cf. antiquus	6.1–7.6	7.0–8.3	7.8–9.2
Symphalangus syndactylus			
subfossilis[1]	7.7–9.7	8.7–10.4	8.2–10.7
S. s. syndactylus	6.4–8.2	6.9–9.7	6.9–9.1

1 HOOIJER, 1960.

Recent Variability and Phyletic Evolution

A. Molar Size

When fossil forms are considered for which an adequate sample from a limited geographical area is available (table V), a tendency for increased size is observed to characterize hylobatid dental evolution from early Miocene until sub-recent times. A comparison of sub-fossil and recent siamang material from Sumatra [KITAHARA-FRISCH, 1971], however, shows this tendency to have been reversed at least since the end of Pleistocene times. Among the extant species of *Hylobates, H. hoolock* of Burma appears to be the one where reduction in size has been least operative. The ranges for M₁ and M₂ are nearly identical with those reported for the *Pliopithecus* sample from Göriach and for *Limnopithecus macinnesi* from Kenya. The Burmese gibbon has also preserved a condition close to the primitive one in two other metrical features, the short length of M₁ relative to M₂ and the only slightly reduced size of M₃ [FRISCH, 1965]. Conversely, the recent forms with the shortest lower molar teeth *(H. lar albimanus* and *Brachytanites*[2]) are characterized by the small difference in length between M₁ and M₂ and the considerably reduced length of M₃.

It would thus seem that reduction in overall size of the dentition, the disappearance of the size sequence characteristic of fossil forms (M₁<M₂<M₃), and the metrical reduction of M₃ are related manifestations of a single evolutionary trend. Of course, the association of all three

2   More commonly, *H. klossi;* cf. I. T. SANDERSON: The Monkey Kingdom, chap. 12, p. 151 (Hanover House, Garden City 1957). [Ed.]

of these phenomena is not always a necessary one; and exceptions can be seen. The Javanese gibbon, for instance, shows a greater difference in size between M_1 and M_2 and less reduction in M_3 than the Bornean form, although they are indistinguishable in the absolute size of their dentitions. When all species and subspecies are considered together, however, the usual association of the three phenomena can be clearly seen and helps in understanding how a number of features characteristic of the evolution of the hylobatid dentition can be viewed as manifestations of a single trend. The same dynamics are likely to hold for dental evolution in other hominoid phyla.[3]

B. Development of the Lingual Cingulum on the Upper Molars

As pointed out previously [FRISCH, 1965], in spite of the variability in the development of this structure within each hominoid genus, less intensive reduction took place in gibbons than in the large apes. Nevertheless, comparison of the living with the fossil forms leaves no doubt that the marked intra- and interspecific variability of this feature in the hylobatid family can be interpreted as the result of a still-operating evolutionary process.

Can something be said about the time when this process began? The marked differences observed between the Javanese and the Bornean gibbons, which have certainly not been separated for more than 400,000 years [KURTÉN, 1960], suggest that reduction of the cingulum is a relatively recent characteristic of hylobatid evolution. Examination of the fossil specimens confirms this indication; no fossil gibbon is known in which the lingual cingulum of the upper molars is absent. It can thus be concluded that the cingulum is a structure with sufficient phylogenetic and taxonomic significance to deserve careful description when assessing the status of fossil forms.

C. Anterior Lower Premolar

The considerable morphological variability observed in the Chiengmai population of *H. lar entelloides* in a tooth commonly considered as one of the stablest of the non-hominid dentition is a striking but very

3   See Dr. KINZEY's article, this volume. [Ed.]

limited phenomenon. The total absence of the triangular-shaped lower anterior premolar in most gibbon species and its occurrence in only certain restricted populations of a single species preclude its insertion in a hypothetical evolutionary trend from sectorial to a hominid molariform type of premolar. Nevertheless, it cannot be denied that this tooth appears much less stable in its morphology than has often been assumed. Taking this instability into consideration makes it less difficult to visualize the derivation of a hominid type of lower anterior premolar from the sectorial type seen in all known Miocene fossil pongids. It is probably an exaggerated notion of the stability of the pongid sectorial P_3 that led some authors [e.g. HUERZELER, 1968] to consider the occurrence of either homorphic or heteromorphic types of P_3 at a given point in the fossil record as sufficient indication of the existence at that time level of entirely separate lines of specialization.

D. Location of the Hypoconulid

More clearly perhaps than the structure of the cingulum, the groove pattern of the lower molars has preserved in gibbons the condition characteristic of early Miocene hominoids. This statement holds especially for the location of the hypoconulid, which has remained central in the majority of first and second molars in a number of species. As was concluded in the case of the reduced condition of the lingual cingulum, the buccal shift of this most distal cusp of the lower molars appears thus to be a rather late evolutionary development, a conclusion again supported by the comparison of Javanese and Bornean gibbons. Separated only in Pleistocene times, these two populations show so much divergence in the configuration of the prevalent cusp pattern, especially on M_1, that it is almost possible to identify the origin of specimens by this feature alone.

Primate Dental Variability and Hominid Phylogeny

On the basis of the evidence examined above, the question remains of how better knowledge of dental variability in primates, particularly in hominoids, may prove helpful in solving problems in hominid phylogeny. One example each will be discussed in the fields of taxonomy and phylogenetic reconstructions.

A taxonomic problem often raised because of its relevance to important evolutionary mechanisms is the adequacy of the so-called *single species hypothesis* [WOLPOFF, 1971a]. For example, is it possible to determine with a fair degree of reliability whether the gracile and robust australopithecines formed truly separate lineages? And, if separate lineages are held to be the case, should the distinction be considered to be specific or generic? It is particularly necessary to consider dental evidence here, since the claim that two australopithecine genera coexisted contemporaneously has relied heavily on differences in the dentition [ROBINSON, 1968]. A part of this controversy in the last ten years has been the fossil find christened by the late L. S. B. Leakey *Homo habilis,* which has become the object of heated discussion. Here again, it is the dentition which forms the focal point of the argument regarding the appropriateness of including this specimen in the genus *Homo* or in *Australopithecus.* Both partisans and opponents of a generic distinction between *Australopithecus* and the specimen called *Homo habilis* argue from dental evidence, particularly metrical data.

Although the evidence considered in the preceding sections cannot be expected to suffice to solve these controversies, some data are presented that should be kept in mind when discussing problems of hominoid taxonomy. It was seen, for instance, that far greater differences in absolute size are found between the molars of two species of *Hylobates (H. hoolock* and *H. lar albimanus)* than between the robust and gracile australopithecines. It might be objected that the two gibbon taxa are widely separated geographically, while the two australopithecine forms appear to have inhabited identical regions of Africa. However, table VI shows that differences greater than those observed in the Transvaal hominids also occur between two hylobatid species with neighboring ranges, *H. hoolock* and *H. lar entelloides.*

The great variability observed in the anterior lower premolar should also, as stressed above, make one wary of relying overmuch on the morphology or dimensions of this tooth in establishing the taxonomic status of a given form. Prudence is all the more in order since the shape of P_3 often makes it difficult for independent investigators to obtain easily comparable measurements. These two factors combined probably explain how LEAKEY [1961] and ROBINSON [1965] could come out with data and conclusions so different when examining the same material attributed by its finder to *Homo habilis.* Thus, the consideration of the data presented above, though not sufficient to decide between the various intepretations given to

Table VI. A comparison of molar lengths in australopithecines and in some hylobatid species

	M¹	M²	M³	M₁	M₂	M₃
A/P¹	90.64	92.26	86.28	95.19	97.21	91.17
Distance	9.36	7.73	13.72	4.80	2.78	8.83
LA/H	83.12	80.89	67.06	82.23	77.60	70.93
Distance	16.88	19.11	32.94	17.77	22.40	29.07
LE/H	86.50	86.33	77.64	89.43	84.06	87.60
Distance	13.50	13.67	22.36	10.57	15.94	12.40

A/P = Molar length in *Australopithecus* expressed as a percentage of the same dimension in *Paranthropus;* LA/H = same index for *Hylobates lar albimanus* and *H. hoolock;* LE/H = same index for *H. lar entelloides* and *H. hoolock.*

1 Measurements of *Paranthropus* and *Australopithecus* are taken from WOLPOFF [1971b].

the hominoid fossil evidence, cautions strongly against attributing too much weight to particular features of the dentition. More specifically, although ROBINSON's [1965] dietary hypothesis may well prove to be correct as further evidence is collected, it is important to realize that dental evidence *by itself* cannot be made to support such an important taxonomic distinction between the two australopithecine lineages. A similar statement is applicable to the respective status of *Australopithecus* and *Homo habilis.*

An application of the hylobatid data developed in this paper to the solution of problems in hominid phylogenetic reconstruction concerns the early hominid form named by Leakey *Kenyapithecus.* When the lower Pliocene deposits of Kenya yielded in 1961 this oldest known African member of the hominid lineage, popular presentations were quick to call it 'the oldest known ancestor of man'. Although few scientists doubt today the hominid status of this form, held by many to be identical to the Siwaliks' *Ramapithecus,* the question of whether or not it should be considered as one of the direct ancestors of *Homo* remains undecided.

As was pointed out elsewhere [FRISCH, 1967b], the absence of a lingual cingulum on the upper molars of the African Pliocene hominid distinguishes them clearly from those of *Australopithecus,* where the cingulum is conspicuous on the second upper molars and present in various degrees of reduction on the first. Complete cingulum reduction, as discussed

earlier, appears to be a rather recent phenomenon and thus probably represents the last stage of an evolutionary trend that in the hylobatid
phylum, at least, has not been reversed for the last twenty million years.
Considering the Pliocene hominid from Kenya as ancestral to *Australopithecus* would, however, imply precisely such a reversal. In the light of
what is known of the history of the cingulum in hylobatid phylogeny and
of its present mode of occurrence, it seems unlikely that *Kenyapithecus*
ought to be regarded as a direct ancestor of *Australopithecus*. And, since
the similarities between *Australopithecus* and *Homo* are close and numerous enough to allow experienced paleoanthropologists to identify a given
specimen as belonging to one or the other genus, removing *Kenyapithecus*
from the ancestry of *Australopithecus* would seem also to disqualify it as
an ancestor of man. Some anthropologists, no doubt, will question whether the presence or absence of such a minute structure as the basal cingulum of the upper molars is a sufficient criterion to exclude a form from the
direct ancestry of *Homo*. The study in depth and across species of the hylobatid dentition strongly suggests, however, that this minute structure is
not to be disregarded.

The controversial questions dealt with in this section are only two of
the many possible applications of the data on dental variability in hylobatids to the analysis of the hominoid fossil record. The legitimacy of such a
comparison between living hylobatid forms and fossil hominoid remains
rests on the observation that of living apes, the gibbons, by virtue of the
morphology of their dentition as well as of other parts of their anatomy,
have apparently remained closest to the Miocene hominoids. It may,
therefore, be assumed that the variability characteristic of their dentition
allows a good estimate of that present in the Miocene forms. Thus, one
may hope that the element of subjectivity always involved in the interpretation of the fossil record can be reduced by such studies and help be giv-
en to those who practice what has been called the difficult 'art of taxonomy'.

Acknowledgments

The author wishes to acknowledge the gracious assistance of Mrs. INGRID PALM
and members of her data processing staff at the Oregon Regional Primate Research
Center for help with the statistical treatment of the data presented in this paper.
Thanks are also due to others at the Center who assisted in the manuscript preparation and especially to Dr. M. R. ZINGESER for his constructive criticisms and for the
time he spent in editing the paper.

References

BIGGERSTAFF, R. H.: Metric and taxonomic variations in the dentitions of two Asian cercopithecoid species: *Macaca mulatta* and *Macaca speciosa*. Amer. J. phys. Anthrop. *24:* 231–238 (1966).

BOOTH, S. N.: Observations on the teeth of the mountain gorilla *(Gorilla gorilla beringei)*. Amer. J. phys. Anthrop. *34:* 85–88 (1971).

FRISCH, J. E.:[4] Dental variability in a population of gibbons *(Hylobates lar)*; in BROTHWELL Dental anthropology, pp. 15–28 (Pergamon, Oxford 1963).

FRISCH, J. E.: Trends in the evolution of the hominoid dentition. Bibl. primat., vol. 3 (Karger, Basel 1965).

FRISCH, J. E.: The gibbons of the Malay peninsula and of Sumatra. Primates *8:* 297–310 (1967a).

FRISCH, J. E.: Remarks on the phyletic position of *Kenyapithecus*. Primates *8:* 121–126 (1967b).

FRISCH, J. E.: The hylobatid dentition; in RUMBAUGH Gibon and siamang (Karger, Basel 1973).

GREGORY, W. K. and HELLMAN, M.: The dentition of *Dryopithecus* and the origin of man. Anthrop. Pap. amer. Mus. nat. Hist. *28:* 1–123 (1926).

GROVES, C. P.: Geographic and individual variation in Bornean gibbons, with remarks on the systematics of the subgenus *Hylobates*. Folia primat. *14:* 139–153 (1971).

HOOIJER, D. A.: Quaternary gibbons from the Malay archipelago. Zool. Verh. *46:* 1–41 (1960).

HORNBECK, P. V. and SWINDLER, D. R.: The morphology of the lower fourth premolar of certain Cercopithecidae. J. dent. Res. *46:* suppl. (1967).

HUERZELER, J.: Questions et réflexions sur l'histoire des anthropomorphes. Annal. Paléont. *54:* 195–233 (1968).

JOLLY, C. J.: The large African monkeys as an adaptive array; in NAPIER and NAPIER Old world monkeys, pp. 139–174 (Academic Press, New York 1970).

KINZEY, W. G.: Evolution of the human canine tooth. Amer. Anthrop. *73:* 680–694 (1971).

KITAHARA-FRISCH, J.:[5] Evolution of the siamang *(Symphalangus syndactylus)* in Southeast Asia during Pleistocene; in BIEGERT Taxonomy, anatomy, reproduction; Proc. 3rd Int. Congr. Primat., Zurich 1970, vol. 1, pp. 67–73 (Karger, Basel 1971).

KURTÉN, B.: Faunal turnover dates for the Pleistocene and late Pliocene. Soc. Scient. fenn. Comment. Biol. *22:* 1–14 (1960).

LEAKEY, L. S. B.: The juvenile mandible from Olduvai. Nature *191:* 417–418 (1961).

NAPIER, J. R. and NAPIER, P. H.: Introduction; in NAPIER and NAPIER Old world monkeys, pp. XI–XVI (Academic Press, New York 1970).

PILBEAM, D. R.: Notes on *Ramapithecus*, the earliest known hominid, and *Dryopithecus*. Amer. J. phys. Anthrop. *25:* 1–6 (1966).

4   See also KITAHARA-FRISCH, J. [Ed.]
5   See also FRISCH, J. E. [Ed.]

ROBINSON, J. T.: *Homo habilis* and the australopithecines. Nature, Lond. *205:* 121–124 (1965).

ROBINSON, J. T.: The origin and adaptive radiation of the australopithecines; in KURTH Evolution and hominisation; 2nd ed., pp. 150–175 (Fisher, Stuttgart 1968).

SIMONS, E. L.: Some fallacies in the study of hominid phylogeny. Science *141:* 879–889 (1963).

SWINDLER, D. R.; GAVAN, J. A. and TURNER, W. M.: Molar tooth size variability in African monkeys. Human Biol. *35:* 104–122 (1963).

WASHBURN, S. L.: An old theory is supported by new evidence and new methods. Amer. Anthrop. *56:* 436–441 (1954).

WOLPOFF, M. H.: Competitive exclusion among lower Pleistocene hominids: the single species hypothesis. Man *6:* 601–614 (1971a).

WOLPOFF, M. H.: Metric trends in hominid dental evolution (Case Western Reserve University, Cleveland, 1971 b).

ZINGESER, M. R.: Odontometric characteristics of the howler monkey *(Alouatta caraya)*. J. dent. Res. *46:* 975–978 (1967).

Author's address: Dr. J. E. KITAHARA-FRISCH, Sophia University, 7 Kioicho, Chiyodaku, *Tokyo* (Japan)

Symp. IVth Int. Congr. Primat., vol. 3: Craniofacial Biology of Primates,
pp. 148–153 (Karger, Basel 1973)

The Position of Proconsul among the Pongids

G. H. R. VON KOENIGSWALD

Senckenberg Museum, Frankfurt am Main

The cingulum is an essential part of the primitive crown pattern of the mammalian dentition. In its original form, it encircles the entire crown, a condition found even in the canines of such Eocene perissodactyla as *Lophiodon* and *Paleotherium*. In its evolution, the cingulum may give rise to cusps, for example the upper molar hypocone and the lower molar hypoconid. It may also give rise to additional ridges, as in the elephants and in certain Suidae. Its presence or absence should be noted since this information may be relevant to the assignment of the taxonomic position of a given species.

This study is primarily concerned with the upper molars of certain Hominoidea. In the upper molars of *Homo sapiens*, the cingulum has completely disappeared, except for an outgrowth on the lingual side mesial to the protocone. This cusp has been called the 'Carabelli cusp' and was formerly regarded as typical for man; but it also can occasionally be observed in *Hylobates* and in Pongidae. It is identical with the 'pericone', first described by STEHLIN in *Pericondon* from the Eocene of Switzerland. This 'Carabelli cusp' also appears in fossil Pongidae and is clearly visible on a molar of *Sivapithecus cf. indicus* from the Lower Pliocene of Chinji, Pakistan [GREGORY *et al.*, 1938, pl. 5, fig. B and C].

In the modern pongids, the inner (or lingual) cingulum (*cingulum internum*) might sometimes be recognized as a thin ledge running around the protocone and culminating in the hypocone; the mesial and distal borders otherwise show no trace of this structure. We might call this type of cingulum formation the 'pongid-type'. The upper third molars of the gorilla are unique since the cingulum may encircle the hypocone [REMANE, 1921].

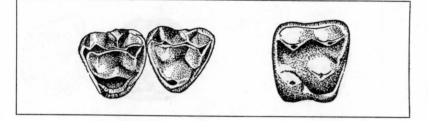

Fig. 1. The cingulum in the upper molars of fossil lemurs. Left: *Megaladapis,* Pleistocene, Madagascar. Triangular molars with cingulum, no hypocone. Right: *Adapis,* Eocene, France. The fourth cusp, the hypocone, is an outgrowth of the cingulum. Not to scale.

A very different type of formation of the inner cingulum, which we might call the 'Proconsul-type', has been described in the Miocene members of the genus *Proconsul* (*P. africanus, P. nyanzae,* and *P. major*). Here a complete inner cingulum can be observed, beginning mesial to the paracone, encircling the entire protocone (and usually also the hypocone), and ending distal to the metacone [LE GROS CLARK and LEAKEY, 1951, fig. 11, 13, 14, 16, 24–28]. The interpretation of this remarkable development is of the greatest importance in determining the proper place of this genus among the Pongidae.

According to LE GROS CLARK and LEAKEY [1951, p. 110]: 'Whether *Proconsul* can be regarded as having any ancestral relationship to the European forms of *Dryopithecus* depends largely on the interpretation of its apparent specializations – particularly the strong and elaborate development of the internal cingulum in the upper molars. It is possible, of course, that, by gradual reduction and eventual disappearance of the internal cingulum, upper molars of the *Dryopithecus* type might have been derived from those of the *Proconsul* type. While such a possibility can hardly be denied, it seems more likely that the tendency towards the elaboration of the cingulum in *Proconsul* is a local specialization of a slightly aberrant type. Such an interpretation would render somewhat less probable the view that this genus could have given origin to *Dryopithecus*.'

SIMONS and PILBEAM [1965] speculate thus upon the meaning of the peculiar cingulum formation in *Proconsul*: 'It is interesting to note that the African species of *Dryopithecus*' (i.e., the three species of *Proconsul*) 'with the exception of the *D. sivalensis* specimens from Rusinga' (the teeth from a species originally referred to as *Sivapithecus africanus*) 'pos-

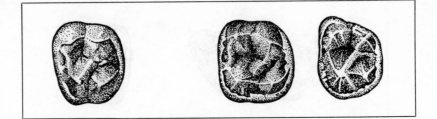

Fig. 2. The cingulum in the upper molars of fossil pongids from the Miocene of East Africa. Left: first upper molar of *'Kenyapithecus'*. Cingulum preserved only at the lower border of the protocone, a condition normal in modern pongids. Right: upper molars of *Proconsul*. A heavy cingulum ('neocingulum') is encircling proto- and hypocone, a highly specialized condition.

sess extensive cingula and complex molar crowns while non-African species have somewhat less distinct or vestigial cingula and relatively simple crown patterns with the exception of *D. fontani*. Does this mean that the African species are descended from a common ancestor, an ancestor not shared by non-African dryopithecines? Are they more primitive ancestors of later, non-African dryopithecines? Have they retained those features while non-African species have lost them? Or are they simply sub-specific variants of wide-ranging species? All these questions await definitive answers and defintive answers require more material' [SIMONS and PILBEAM, 1965, p. 140].

However, the question (thought by LE GROS CLARK and LEAKEY to be impossible to settle conclusively) of the possibility of the persistence to a much later date of *Proconsul* representations, which was raised by LÖNNBERG's [1937] discovery of several teeth of a large ape (named by him *Proconsuloides naivashae*) in Gamlian deposits southeast of Lake Naivasha, can be answered. It is certain that LÖNNBERG's material (a canine and two incisors) does not belong to a primate, but to a carnivore, probably of the genus *Felis; Proconsul* thus remains limited to Miocene Africa, specifically East Africa.

To understand fully the meaning and importance of the Proconsul type of cingulum, we must go back to certain lemuroid forms, where a fourth cusp has not yet developed and the upper molars are still triangular (fig. 1). On the buccal side the row of molars (and premolars) forms an uninterrupted wall, but on the lingual side there are typical triangular gaps between the teeth. The inner cingulum is generally well developed as a low rim. Only when the hypocone as an outgrowth of the cingulum distal to the protocone has been added, have the gaps closed. The molars are now better adapted to chewing, and the teeth now can do their work more efficiently. When the outline of the molars changes from the primitive triangular to the more advanced quadrangular type, an initial stage can be expected in which the expanded molars come just into contact with each other without suppressing the original cingula. This is exactly the stage shown by the molars of *Proconsul*. The cingulum is low (and sometimes beaded); and an anterior groove (fovea anterior), typical for *Dryopithecus* and most Pongids, has not yet been differentiated because the mesial wall is not yet high enough (fig. 2).

The conditions in two specimens of *Pliopithecus* (*Epipliopithecus*) *vindobonensis* Zapfe from Neudorf are most intriguing. While there is no trace of a cingulum in individual II (regarded as a female) in the first two upper molars, individual I (considered to be a male) has a broad and beaded cingulum that surrounds the hypocone, as in *Proconsul*. But this 'neocingulum' in the first and second molars is confined to the lingual border [ZAPFE, 1960, fig. 5]. Only in the third molar is the distal border also included. The cingulum serves to enlarge the chewing surface; it fades out at the mesial border of all molars and at the distal border of the lower third molars. This condition contrasts with that in *Proconsul*, where the cingulum is also prominent between the molars. The strong cingulum of *Proconsul* is in itself primitive and consequently might be considered to represent an expected stage in the evolution of the Hominoidea. (The apparent inflation of the cingulum along the lingual border and the beading of this structure might then be regarded as an overspecialization.)

If the foregoing analysis is correct, then *Proconsul* should indeed be excluded from the ancestry of the later pongids. That the type of the modern pongid molar already existed at the same time as Proconsul lived is evident from the species originally called by LE GROS CLARK and LEAKEY [1951, fig. 41–45] '*Sivapithecus africanus*', which coexisted with *Proconsul*. One first upper molar of *S. africanus* [cf. their fig. 44] even shows an incipient ridge between metacone and hypocone typical of certain *Dryopithecinae;* and the fovea posterior is already well defined.

The coexistence in the Miocene of two forms representing two types of dental evolution makes it difficult to accept the opinion that a direct linear evolutionary relationship could exist between *Proconsul* and the modern pongids. The differences between the Miocene *Proconsul* and recent species are too great. The suggested ancestral relationship of the large *Proconsul major* and the modern gorilla was probably proposed because of their similar proportions. The same objection applies to the orang, which according to ANDREWS [1970] should be considered to be related to one of the East African fossils. The modern orang is difficult to evaluate, as his type has apparently been influenced and changed by isolation on Borneo and Sumatra. Fossil orangs, however, from China and Indochina have been of gorilla size.

In the eurasiatic Miocene pongids, parts of the lateral cingula between the molars have survived, but not enough unworn teeth have been found to permit adequate study. The Pliocene pongids do not differ from the modern ones, whose lateral cingula have been incorporated in the outer wall due to the expanding cusps; this incorporation is less evident in the Pliocene gorilla than in the orang [REMANE, 1921, fig. 25]. In the Upper Tertiary, the Proconsul-type of cingulum seems to be limited to *Proconsul* proper; it is highly characteristic for this genus and thus allows us to distinguish *Proconsul* from other contemporaneous pongids.

Summary

The highly developed upper molar cingulum of *Proconsul*, a fossil pongid from the Miocene of Kenya, is not an overspecialization but rather a primitive condition. Other pongids found in the same geologic stratum resemble recent species in that the cingulum is incorporated into the outer wall of the tooth crown. It is therefore unlikely that *Proconsul* is ancestral to any of the living pongids.

References

ANDREWS, P.: Two new fossil primates from the Lower Miocene of Kenya. Nature, Lond. *228:* 537–540 (1970).

GREGORY, W. K.; HELLMAN, M. and LEWIS, G. E.: Fossil anthropoids of the Yale-Cambridge India Expedition of 1935. Carnegie Inst. Washington Publ. No. 495, pp. 1–27 (1938).

LEGROS CLARK, W. E. and LEAKEY, L. S. B.: The Miocene hominoidea of East Africa. British Museum (Nat. Hist.), Fossil Mammals of Africa, No. 1, pp. 1–117 (1951).

LÖNNBERG, E.: On some fossil mammalian remains from East Africa. Ark. Zool., Upsala *29-A:* 1–23 (1937).

REMANE, A.: Beiträge zur Morphologie des Anthropoidengebisses. Wiegmanns Arch. Naturgesch. Abt. A *87:* 1–179 (1921).

REMANE, A.: Zähne und Gebiss (der Primaten); in Primatologia III, pp. 637–846 (Karger, Basel 1960).

SIMONS, E. L. and PILBEAM, D. R.: Preliminary revision of the Dryopithecinae, Pongidae, Anthropoidea. Folia Primat. *3:* 81–152 (1965).

ZAPFE, H.: Die Primatenfunde von Neudorf a.d. March, Tschechoslowakei. Schweiz. Pal. Abh. *78:* 1–293 (1960).

Author's address: Dr. G. H. R. VON KOENIGSWALD, Senckenberg-Museum, Senckenberganlage 25, *D–6 Frankfurt am Main 1* (FRG)

Symp. IVth Int. Congr. Primat., vol. 3: Craniofacial Biology of Primates,
pp. 154–179 (Karger, Basel 1973)

Dental Variability in the Tree Shrews (Tupaiidae)

D. G. STEELE

University of Alberta, Edmonton

Introduction

In recent years, there has been renewed interest in the tree shrews
(family Tupaiidae) especially concerning their behavior [ELLIOT, 1969;
MARTIN, 1966, 1968; SORENSON and CONWAY, 1964, 1966; VANDEN-
BERGH, 1963], ecology [ELLIOT, 1969], and, in particular, their classification. Undoubtedly, the most controversial issue regarding tupaiid taxonomy has revolved around the question of whether they should be included
within the order Insectivora or the order Primates, or possibly classified
as a separate order. Although initially classified as insectivores, CARLSSON
[1922], LE GROS CLARK [1924, 1925], GREGORY [1910] and others began
to emphasize the similarities between the tree shrews and the Prosimii. As
a consequence of their work and of SIMPSON's [1931] description of the
early Tertiary *Anagale* as a tupaioid, many later researchers accepted the
classification of the Tupaiidae as primates [SIMPSON, 1935, 1945; NAPIER
and NAPIER, 1967]. However, with BOHLIN's [1951] and later McKENNA's
[1963] dismissal of *Anagale* from the superfamily Tupaioidea, the question of the ordinal position of the tree shrews has once again been raised.
Of particular note in the controversy have been CAMPBELL's [1965,
1966], HILL's [1965], LUCKETT's [1967], MARTIN's [1966, 1968],
McKENNA's [1963, 1966], SZALAY's [1968], and VAN VALEN's [1965] arguments for exclusion of the tree shrews from the primates and MEISTER
and DAVIS's [1956] and SABAN's [1956, 1957] support of their inclusion.

As the history of the controversy over the allocation of the tree
shrews to an order is indicative of our meager knowledge of these crea-
tures, so is the history of the classification of the tupaiids at the specific

and generic level. After DIARD's initial description of *Tupaia glis* in 1820, the first phase of classifying the tree shrews was dominated by the description of newly-discovered forms. In fact, this phase of tupaiid classification has continued more or less intermittently up to the present, the most recent [1962] being DAVIS's description of *Tupaia (Lyonogale) tana kretami*. During these 140-odd years of naming tree shrews, more than 100 distinctive forms have been recognized. The only researcher to review carefully the entire family was M. W. LYON in 1913. At that time, LYON described two subfamilies, the monogeneric Ptilocercinae and the Tupaiinae, and six genera, *Anathana* and *Lyonogale* [see CONISBEE's 1953 index of mammalian genera for the change of the name from LYON's *Tana* to *Lyonogale*] being so defined for the first time. For these six genera, three species were described for *Anathana*, three for *Dendrogale*, one for *Ptilocercus*, one for *Urogale*, seven for *Lyonogale*, and 31 for *Tupaia*.

Since LYON's revision of the family, several of his conclusions have proved controversial. Initially, LYON defined 31 species of *Tupaia*, many found in Indochina and on the adjacent islands. The Malaysian peninsular forms were divided into three species, *T. glis*, *T. lacernata*, and *T. belangeri*. Shortly after LYON's revision, an additional form, *T. clarissa*, which occurred between the populations of *T. lacernata* and *T. belangeri*, was described. KLOSS [1918], on the basis of the gradation in form and pelt characteristics, placed all four species within *T. glis* but recognized their differences on the subspecific level. CHASEN [1940], following KLOSS' lead, classified the majority of the remaining portion of LYON's species within *T. glis*. The only justification CHASEN gave for these changes of LYON's classification was the allopatric distribution of the majority of LYON's species of *Tupaia* and the statement [p. 7]:

'... I can find no reason for keeping them apart. Even the subspecies most remote from each other in color can be linked by intermediate forms and any grouping on general facies is at variance with a geographical arrangement.'

In addition to these changes, CHASEN [1940] also felt that the species *T. splendidula* and *T. carimatae* would subsequently be shown to be members of the species *T. glis*. These revisions have resulted in the reduction of LYON's 31 species to 11 recognized as valid by CHASEN [1940] or ELLERMAN and MORRISON-SCOTT [1951].[1]

Since the publication of CHASEN's list, MEDWAY [1961] has reaffirmed LYON's classification of *T. splendidula*. On the basis of three new

1   NAPIER and NAPIER [1967] list 12 species of Tupaia. [Ed.]

skins collected from West Kutai, central east Borneo (Kalimantan), MED-WAY has been able to demonstrate the existence of *T. splendidula* and its sympatric relationship with *T. glis longipes*. MEDWAY also stated in his report that three other forms, *T. carimatae*, *T. g. lucida*, and *T. g. natunae*, should now be considered subspecies of *T. splendidula*, a classification used by NAPIER and NAPIER [1967], whose taxonomy is followed throughout this paper. MARTIN [1968] has also questioned CHASEN's and ELLERMAN and MORRISON-SCOTT's lumping of several taxa within *T. glis*, suggesting that the different number of mammae and the possible differences in litter size between the southern peninsular forms of *T. glis* and the more northerly forms, LYON's *T. belangeri/chinensis* group,[2] may indicate the existence of two distinct species.

The primary taxonomic problem pertaining to the genus *Tupaia* then is whether the many allopatric populations from the mainland of Malaysia and surrounding islands should be grouped into one or more species. CHASEN [1940] and ELLERMAN and MORRISON-SCOTT [1951] seemed to feel that the various allopatric populations were subspecies of the highly varied and ubiquitous *T. glis*; while MEDWAY [1961] and MARTIN [1968] felt that *T. glis*, as defined by those authors, represents two or more valid species. MAYR [1963] has pointed out that, while allopatric distributions of morphologically similar populations are indicative of conspecificity, they are not an infallible guide. Allopatry as an indicator of conspecificity is a working hypothesis, a hypothesis which should be tested in each case [MAYR, 1963].

Two other taxonomic assessments by LYON which have proved controversial are the validity of his genera *Lyonogale* (originally *Tana*) and *Anathana* and the distinction of the two subfamilies. The terrestrial tree shrew *Tupaia (Lyonogale) tana* was placed in a separate genus from *Tupaia* in 1913 by LYON, and subsequently reallocated to *Tupaia* by CHASEN [1940]. Since then, ELLERMAN and MORRISON-SCOTT [1951] have proposed the subgeneric distinction *Lyonogale* for the species; and this classification has been accepted by MEDWAY [1965] and NAPIER and NAPIER [1967].[3] In a similar fashion, FIEDLER [1965] has questioned the generic distinction of *Anathana*, preferring to reallocate it to the genus *Tupaia*; the NAPIERS, however, have retained the separate generic classification.

2   The NAPIERS have retained these two forms as separate subspecies of *T. glis* [p. 343]. [Ed.]
3   The NAPIERS list an additional species, *T. dorsalis,* in this subgenus [p. 344]. [Ed.]

The only author to question Lyon's classification of the Tupaiidae at the level of the subfamily has been Davis [1938], who suggested that the intermediate morphological nature of *Dendrogale*, a member of the subfamily Tupaiinae, is cause to question Lyon's dichotomous classification.

In spite of this interest in the taxonomy of the family, Hill's [1960] demonstration of a clinal variation in coat color in the Malay peninsular tree shrews is one of the few studies attempting to describe and quantify the range of variation of some of the features of the tupaiids. The lack of demonstrated variability of morphological features is particularly apparent when the dentition of the tree shrews is considered. Mivart [1867], Gregory [1910], and Lyon [1913] presented descriptions of tupaiid dentition; but little variability was considered. Other than these generalized descriptions, the only other work pertaining to tupaiid dentition with which I am familiar is Kindahl's [1957] research on tooth development in *T. javanica*. The need for studies of variability, particularly dental variability, within the tupaiids becomes acutely apparent when one realizes that the major part of early Tertiary insectivore and primate material preserved consists of teeth. The fact that both *Adapisoriculus* [van Valen, 1965] and *Meselina* [Szalay, 1968, 1969] have been suggested as possible tupaiids on the basis of dentition alone underscores the need for more detailed research on the dentition of the tree shrews.

The purpose of the present study, therefore, is to describe briefly the dentition of the family Tupaiidae and to document some of the variation encountered in the 301 specimens observed. This examination is in three parts. First, a descriptive report of the dentition and the dental variability is presented. A Q-mode cluster analysis of the tree shrews based upon 43 discrete dental characteristics follows the qualitative description of the variation observed. The third part presents an R-mode cluster analysis of these traits and relates them to the specimens and groups of specimens. It is hoped that the data presented and so analyzed will demonstrate the range of dental variability present in the teeth of tree shrews, help clarify certain areas of controversy, and indicate new areas of taxonomic interest.

Sample and Nomenclature Used

The major portion of the tree shrew sample used in this study is contained in the collections of the US National Museum, Smithsonian Institution, Washington,

Table I. List of the 14 species of the Tupaiidae examined (with sample size of each) in the present study [classification follows Napier and Napier, 1967]

Name	General locality	Male	Female
Subfamily Ptilocercinae			
Genus *Ptilocercus*			
Ptilocercus lowii	Malaya	3	–
Subfamily Tupaiinae			
Genus *Dendrogale*			
Dendrogale melanura	North Borneo	4	1
Dendrogale frenata	Thailand and South Vietnam	6	3
Genus *Tupaia*			
Subgenus *Tupaia*			
Tupaia glis			
T. g. batamana	Batam Island	3	4
T. g. cambodiana	Thailand	8	10
T. g. chrysogaster	Mentawi Islands	2	9
T. g. clarissa	Peninsular Thailand and Mergui Archipelago	13	10
T. g. demissa	Sumatra	1	–
T. g. discolor	Banka Island	5	4
T. g. ferruginea	Malaya	9	17
T. g. jacki	Western Sumatra	–	1
T. g. kohtauensis	Koh Tao Island	2	3
T. g. lacernata	Pulo Langkawi and Terutau	1	3
T. g. laotum	Thailand	1	–
T. g. lepcha	India (Bengal) and Nepal	1	1
T. g. modesta	Hainan Island	1	–
T. g. olivacea	Thailand	5	4
T. g. phaeniura	North Sumatra	1	2
T. g. phaeura	Singkep Island	–	2
T. g. salatana	Borneo	3	–
T. g. siaca	Sumatra	2	1
T. g. siccata	India (Manipur)	–	1
T. g. sinus	Koh Chang Island	1	1
T. g. sordida	Tioman Island	1	4
T. g. ultima	Koh Pangan Island	–	1
T. g. versurae	Northern Burma, Eastern India, and Southern China	2	3
T. g. wilkinsoni	Peninsular Thailand	4	1
Tupaia gracilis			
T. g. gracilis	Borneo	2	2
T. g. inflata	Banka and Billiton Islands	2	–
Tupaia javanica			
T. j. occidentalis	Java	3	5

Table I (continued)

Name	General locality	Male	Female
Tupaia minor			
T. m. malaccana	Eastern Sumatra and Malaya	5	4
T. m. minor	Borneo	7	5
Tupaia montana			
T. m. baluensis	Sabah, Mt. Kinabalu	20	13
Tupaia mulleri	Borneo	1	1
Tupaia nicobarica			
T. n. nicobarica	Greater Nicobar Island	5	6
T. n. surda	Lesser Nicobar Island	3	4
Tupaia picta	Sarawak	2	–
Tupaia splendidula			
T. s. carimatae	Karimata Islands	–	1
T. s. splendidula	Borneo	1	1
Subgenus *Lyonogale*			
Tupaia tana			
T. t. bunoae	Big Tambelan Island	1	1
T. t. chrysura	Borneo	1	–
T. t. paitana	Borneo	–	1
T. t. sirhassensis	Natuna Islands: Sirhassen	1	2
T. t. tana	Borneo and Sumatra	17	15
T. t. utara	Borneo	1	–
Genus *Urogale*			
Urogale everetti	Philippine Islands: Mindanao	2	1

D. C. In addition, specimens of six of the subspecies of *T. (Lyonogale) tana* and of *Urogale everetti* from the American Museum of Natural History, New York, were examined. Table I lists by taxonomic category the 301 specimens examined and gives the sample number for each of the 46 taxa represented and the locality for each taxon. As stated previously, the classification used in this paper follows, with little exception, the systematic list of NAPIER and NAPIER [1967], although not all the forms listed by the two authors are represented in the sample.[4]

Unavoidably, the detailed description of teeth requires a specialized vocabulary, which may vary from author to author. Figures 1 and 2, therefore, illustrate a typical upper and lower molar with the features referred to in this paper labeled according to the nomenclature of SZALAY [1969]. Following VAN VALEN [1966], SZALAY's terms will be used to describe features on premolars as well as molars.

4   For example, no specimens from the monospecific genus *Anathana* were examined or from the species *T. dorsalis* or *T. palawanensis*; similarly, many of the listed subspecies of *T. glis*, *T. tana*, and a few of those of other species, are not represented. In addition, the species of *Dendrogale* cited do not correspond with the ones given by the NAPIERS. [Ed.]

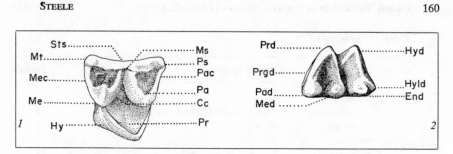

Fig. 1. A representative example of an upper second molar from *Tupaia glis* with the characteristic features identified: Cc = centrocrista (= postparacrista plus premetacrista referred to in text, section D, p. 165); Hy =hypocone; Me = metacone; Mec = metacrista; Ms = mesostyle; Mt = metastyle; Pa = paracone; Pac = paracrista; Pr = protocone; Ps = parastyle; Sts = stylar shelf.

Fig. 2. A representative example of a lower second molar from *Tupaia glis* with the characteristic features identified: End = entoconid; Hyd = hypoconid; Hyld = hypoconulid; Med = metaconid; Pad = paraconid; Prd = protoconid; Prgd = precingulid.

The Dentition of the Tupaiidae; Variability among the Taxa

A. The Dental Arcade as a Complete Complex

The dental formulae for the Tupaiidae is $\frac{2133}{3133}$. If a $\frac{3143}{3143}$ dental formula for the primitive Eutherian condition is assumed, the Tupaiidae have lost one upper incisor and one upper and lower premolar. Since the general condition of the two remaining maxillary incisors is for the first to be the larger, it is assumed that the third incisor (still smaller than the first and second in the primitive state) has been lost [LYON, 1913]. This hypothesis is supported by the fact that the lower third incisor is generally diminutive, actually functioning as an incisor in only a portion of the Tupaiidae. Also, of the 301 specimens examined, the only congenitally absent incisor has been a lower third incisor. Conversely, both upper and lower premolars, in both primitive and modern forms, increase in size from front to back, an order suggesting that the lost premolars were the first ones [LYON, 1913].

Apparently, the dental complex of the tupaiidae functions in four ways. The upper incisors, canines, and, to a lesser extent, the first premolars have a grasping and stabbing function. Generally, these teeth are styliform with little differentiation in size between the teeth. *U. everetti*, however, has an enlarged and fang-like lower canine and a fang-like upper second incisor that functions as a canine. The premolars and the molars also function sectorially, cutting and shearing food; and the molars and, to a certain extent, the last premolars, function in a grinding fashion.[5] The fourth function of the dentition is in grooming; and possibly for this purpose the lower incisors, and to a certain degree the canines, are procumbent.

B. Incisors and Canines

In discussing the dentition, I shall consider the two subfamilies separately. The upper incisors, as well as the canine and the first upper premolar of the Tupaiinae tend to be styliform. The major morphological variations noted in these teeth are: (1) differences in the relative size of the teeth, (2) differences in the degree of mesial reflection of the first incisors, and (3) differences in the width of the diastemata separating the teeth.

Generally, the upper first incisors tend to be larger than or equal to the second incisors, although those of *Urogale* are an exception to this generalization. However, even among those species in which the generalization holds true, the actual size differences between the two incisors vary between populations. Differences in the mesial reflection of the first and second incisors seem to accompany a mechanical function related to the size of the nostril opening. In the tupaiids, the nostril opening is approximately trapezoidal, with the smaller of the parallel borders forming the base of the opening. The roots of the first incisors, just lateral to the opening, follow the sloping sides of the nostrils. The upper incisors are separated from each other and from the canine by diastemata. The larger diastema separates the second incisor from the canine, and it is distinctly wider in those species characterized by elongated rostra *(T. tana* and *U. everetti).*

The lower incisors, because of their modified nature in response to the function of grooming, are much more differentiated from the lower canines and first premolars than are the upper incisors from the upper canines and premolars. In general, the first two lower incisors are long and narrow, with the second incisor being the larger. In cross-section, these incisors are flattened mesiodistally and have a thickened longitudinal ridge on the lingual aspect of the tooth. The third incisor is distinctly

5   Cf. HIIEMÄE and KAY, this volume. [Ed.]

smaller than the other two, its size varying between less than one-half to slightly more than one-half the length of the second incisor. Although this tooth also is flattened mesiodistally, it generally lacks the thickened longitudinal ridge on the lingual surface.

The upper canine is part of a tooth series consisting of the incisors, canine, and the first premolar, all of which tend to be caniniform. Generally, these four teeth range in order of decreasing size from the first incisor to the first premolar. The one very obvious exception to this generalization is *Urogale,* in which the second incisor is the largest tooth in this series, the canine being diminutive in comparison. In *T. tana* and in *T. glis salatana,* similar in facial appearances to *T. tana,* the incisors, canine, and first premolar are similar in size and are small in relationship to the elongated rostrum. In *Dendrogale frenata* and *D. melanura,* the canine has a broader base relative to the length than is typical for the Tupaiinae, and often there is an accessory cusp on the posterior aspect of the tooth. This is the only genus within this subfamily that exhibited a double-rooted canine. All eight specimens of *D. frenata* examined have a double-rooted canine, as do five of the seven specimens of *D. melanura.*

The general form of the upper canine is a simple cone greater in size than the first premolar, but smaller than the second incisor. Three possible exceptions to this generalization were noticed. In *T. g. demissa, T. g. discolor,* and *T. g. jacki,* the canines seem to be relatively smaller than is common in the other subspecies of *T. glis;* the base of the canine is also broader in relationship to height than is typical for the species. Unfortunately, this observation was based on only one specimen each of *T. g. demissa* and *T. g. jacki,* in addition to the nine specimens of *T. g. discolor* examined. It is interesting to note that all three forms are from Sumatra: *T. g. demissa* from Northeast Sumatra, *T. g. discolor* from the island of Banka just east of Sumatra, and *T. g. jacki* from Western Sumatra around Tapanuli Bay.

The lower canines in the Tupaiinae vary among the species both in relative size and degree of procumbency. In *T. montana, T. nicobarica,* and *U. everetti,* the lower canines are much larger than in the other species of tree shrews, exceeding the second incisor in length. In general appearance, the canines in these three species are fang-like and quite distinct in shape from the lower canines of the rest of the tupaiids. The only sexually-dimorphic size difference in the lower canines was found in *T. montana.* Of the eight adult males and four adult females available, all could be correctly identified as to sex on the basis of canine size alone. If a sex difference in canine size exists in *T. montana* and if this sexually-dimorphic trait occurs only in this species, it should prove interesting to look for behavior patterns in the species that may be correlated with the sexual dimorphism. The only other sexually-dimorphic trait that has been recognized in the tupaiid skull is a difference in interorbital breadth; it has been reported to date only for *T. g. belangeri* [MARTIN, 1968].

The upper incisors and canine of the monospecific *Ptilocercus* (the only genus of the subfamily Ptilocercinae) are quite different from those of the Tupaiinae in several respects: (1) the upper incisors are flattened in a mesiodistal plane, an atypical condition for the Tupaiinae; (2) the diastema between the second upper incisor and the canine is greatly reduced in *Ptilocercus*; and (3) the upper canine is premolariform, being double-rooted and virtually indistinguishable from the upper first premolar. The lower incisors and canines of the two subfamilies are equally distinctive. Although the first two incisors of *Ptilocercus* are as procumbent as the incisors of the Tupaiinae, they are not as long. The third lower incisor of *Ptilocercus* is virtually nonfunctional, being little more than a peg, similar in conformation to the first lower premolar. Because of the limited sample of *Ptilocercus* available, nothing can be said of the dental variation within *P. lowii*.

C. Premolars

The upper premolars of the Tupaiinae range in complexity from a simple, single-cusped tooth (P^2) to a two-cusped tooth (P^4), which in some respects may function as part of the molar row. Generally, the anterior premolar (P^2) can be characterized as a single cusped tooth elongated in a mesiodistal plane. The intermediate premolar (P^3) is larger in size than the anterior premolar and has one large cone (paracone), which may or may not be bordered by paracristae on the mesial and distal borders. In some groups, there may be an expansion of the paracristae into a minor cusp (protocone) on the lingual border. The buccal surface of P^3 is generally smooth, although a slight depression on the posterior buccal surface of the crown may be present. The posterior premolar (P^4) is the largest of the premolars and is also the most complex in structure. The paracone again dominates the tooth and is bordered mesially and distally by paracristae. On the lingual aspect of the tooth, the protocone is typically quite prominent. The buccal aspect of the tooth is generally characterized by a weakly-formed stylar shelf, which may or may not be complete.

The only variation noted in P² among the Tupaiinae is a relative difference in size. In *T. gracilis, T. minor, D. frenata,* and *D. melanura,* P² is relatively smaller than it is in the rest of the Tupaiinae. It is interesting to note that the relatively smaller P² happens to occur in three of the smallest tree shrews.

P³ is the most variable tooth in the tupaiids. Although generally having three roots, double-rooted second premolars were observed in specimens of *T. minor* and *T. nicobarica.* As previously noted, the paracone is usually bordered mesially and distally by paracristae; but in *T. javanica, T. mulleri, T. minor, T. picta, T. nicobarica, T. (Lyonogale) tana,* and *U. everetti,* the paracristae are either weakly developed or not present. The occurrence and size of the protocone on P³ is equally variable, being poorly developed or non-existent in the species just listed as well as in *T. gracilis, T. montana,* and various subspecies of *T. glis.* In fact, a well-developed protocone on P³ is typical only for the genus *Dendrogale.*

P⁴ is much more stable in morphology throughout the Tupaiinae. The major variation noted in this tooth is the degree of development of the stylar shelf. In *T. mulleri, T. minor,* and *T. nicobarica,* the stylar shelf is weakly developed and discontinuous along the buccal surface. In *T. glis* and *T. tana,* the stylar shelf is variably developed. Of particular note is the configuration of P⁴ in *Dendrogale.* In this genus, the paracone is quite high, but is relatively narrower than is typical for the Tupaiinae; and the protocone is larger than in the other Tupaiinae.

The lower premolars, like the upper premolars, increase in size and complexity from front to back. P_2 is a small, single-cusped tooth with one root. P_3 is larger in size than P_2 and has two roots and generally one prominent cusp; in some groups the paraconid and a talonid may be incipiently developed in this tooth. P_4 is the largest of the series and the most molariform; the trigonid has a large protoconid, and the paraconid and metaconid cusps are well differentiated but variable in size. The talonid on the last premolar tends to be small but well differentiated from the trigonid; it is approximately one-half the length and width of the trigonid. In this respect, the last premolar is quite distinct from the first molar, which has a talonid as large or larger than the trigonid.

P_2 as mentioned above, is quite small; and little variation other than size was noticed in this tooth. P_3, on the other hand, is most variable, corresponding in this respect to P³. A small paraconid is present on P_3 in some specimens of *T. gracilis, T. splendidula,* and *U. everetti;* while a small metaconid was observed only on *T. s. carimatae.* The talonid basin is only incipiently developed on all specimens, except *T. montana* and *T. gracilis inflata.* On P_4 the paraconid seems to be well differentiated on all specimens examined. The metaconid is also well differentiated on all except *T. gracilis.* The talonid basin is also present on all specimens, but in *T. montana* and *T. picta* it is slightly smaller than in the other species.

The upper premolars of *Ptilocercus* are quite distinct from those of the Tupaiinae, except for P^2, which is remarkably similar in the two subfamilies. P^3 in *Ptilocercus* is quite small, approximating the size of P^2; in this respect, *Ptilocercus* is markedly different from the Tupaiinae. In the latter group, P^3 approximates in size the larger P^4, rather than P^2. P^4 of *Ptilocercus* is unique in having a prominent hypocone, as well as a protocone.

In *Ptilocercus,* the two anterior lower premolars (P_2, P_3) are simple single-cusped teeth; P_2 is larger than P_3 and almost indistinguishable from the canine. In contrast, the premolars in the Tupaiinae increase in size mesiodistally. P_4 of *Ptilocercus,* on the other hand, is molariform, with three cusps distinguishable on the trigonid and an incipiently-developed talonid.

D. Molars

In the Tupaiinae, the upper molars with the exception of the third tend to be quadrate in shape and to decrease in size from the first molar to the last. The three prominent cusps on each tooth are the paracone, metacone, and protocone. In addition to these major cusps, there is a stylar shelf on the buccal side of the tooth containing three small styles, the parastyle, the mesostyle, and the metastyle. On the distolingual aspect of the first and second molars, a small hypocone may be present; when present, it is dwarfed by the protocone. The three major cusps of the tooth are connected by crests. Diverging from the paracone are the preparacrista and postparacrista. Similarly, the corresponding crests diverging from the metacone have been termed the premetacrista and postmetacrista and the crests diverging from the protocone, the preprotocrista and postprotocrista. Although the third molar is generally similar to the first two molars, it is much smaller; and the metacone on this tooth is also much smaller. No hypocone was observed on a third molar.

Dental variation observed in the upper molars, other than in size, involved the development of the hypocone on the first and second molars, the degree of development of the metacone on the third molar, and whether or not the mesostyle on the stylar shelf is bifid. In general, the hypocone is variably developed among the species of *Tupaia,* ranging from a well-defined cusp to nothing more than a swelling of the protocone in the general region where the hypocone is usually developed. Molars of *Dendrogale, T. minor,* and *T. gracilis inflata* lack hypocones. The hypocones on the few specimens of *Urogale* examined are very well developed; in fact, in this species, the hypocone is the most prominent of all the Tupaiinae. The mesostyle, bifid in the majority of specimens examined, is single only in *T. minor* and *T. gracilis.* The size of the metacone on the third molar seems to be highly correlated with the overall size of the skull, the cusp being smallest in the smaller species, *T. minor, T. gracilis,* and *Dendrogale.*

In the Tupaiinae, the lower molars decrease in size posteriorly, with the third molar being distinctly smaller than the first two. The paraconid, metaconid, and protoconid are quite prominent on the trigonid and are approximately one-half again as tall as the cusps on the talonid. The talonid basin is rimmed by the hypoconid, hypoconulid (sometimes absent on the third molar), and entoconid cusps. In addition, there may or may not be a distinct entoconulid cusp on the first and second molars of *Tupaia nicobarica.* A precingulum is present on the mesial border of the third and second molars of some of the tupaiid species, but it was never observed on the first molar. When present, the precingulum is generally most pronounced on the third molar.

The morphology of the lower molars is remarkably similar throughout the Tupaiinae, with only three features varying to a noticeable degree: (1) the size of the talonid on the third molar, (2) the development of the hypoconulid on the third molar, and (3) the development of precingula on the second and third molars. The first two traits are positively correlated with each other and with the overall size of the tree shrew. In the small tree shrews, *T. gracilis, T. minor,* and *Dendrogale,* the talonid is small and the hypoconulid absent. On the other hand, in *Urogale,* the largest species, the talonid is the most developed; and the hypoconulid is present. Precingula are entirely lacking only in *Urogale* and are restricted to the last molar only in *T. nicobarica* and *T. montana.* In the other species, precingula are developed to a variable degree on both the second and the third molars.

Although Lyon [1913] felt that the dentition of *Ptilocercus* was not very distinctive, he nevertheless described several features of the upper molars that distinguish the two subfamilies. The cusps on the molars of *Ptilocercus* tend to be much more bunodont than they are in the Tupaiinae. Correspondingly, the crests on the teeth of *Ptilocercus* are more weakly defined. The hypocone on all three *Ptilocercus* molars is very well developed and distinctive, being derived from an internal cingulum. Lyon's [1913] figures 5 and 15 admirably illustrate the differences between the hypocones in the two subfamilies of Tupaiidae. Internal cingula are present on the mesial and distal aspects of all of the *Ptilocercus* molars, but are absent in the Tupaiinae. Another characteristic recognized by Lyon as unique to *Ptilocercus* is the lack of a mesostyle on the stylar shelf.

The lower molars of *Ptilocercus* are readily distinguishable from the lower molars of the Tupaiinae only by a well-defined external cingulum present on the three molars of *Ptilocercus*. In addition, the closely approximated molars of *Ptilocercus* make it difficult to distinguish the hypoconulid, which is overhung by the larger paraconid of the molar posterior to it.

Statistical Analysis of the Data

A. Cluster Analyses

In an attempt to evaluate more objectively the dental variation observed, 49 specimens were selected by the author; and for each specimen, 43 discrete traits were coded and recorded. Q-mode and R-mode analyses were then computed using the technique of simple matching coefficients (SMC) [Sokal and Michener, 1958; Sokal and Sneath, 1963]. The technique involves the comparison of all possible pairs on the basis of the number of matches (m) of each variable, divided by the total number of matches possible (n): m/n = SMC. The Q-mode cluster analysis thus resulted in 1,176 possible pairs of tupaiids compared on the basis of the 43 traits; and the R-mode analysis resulted in 903 possible pairs of traits compared on the basis of 49 animals. The actual unweighted pair group method of cluster analysis used by the author is a modification of Computer Contribution 38 of the State Geological Survey of the University of Kansas, Lawrence [Weishart, 1969]. The modification entailed adjusting the SMC to accommodate variables with three character states.

The 43 discrete traits utilized in the cluster analyses are presented in table II. Traits 1–6 define the relative size of the upper and lower incisors, the upper canine, and the upper and lower first premolars. The relative size of the other teeth did not seem to vary noticeably among the taxa considered. Traits 12–15 define the root number for those teeth that varied in the number of roots present. The remaining traits define morphological features of the teeth. Five traits (1, 7, 11, 25, and 28) were divided into three character states, coded as '0', '1', or '2'; the others were divided into two character states, coded as '0' or '1'.

For analysis, one specimen from each of the available taxa was used (with two exceptions). The taxon *T. glis* is thus represented, for instance, by one specimen from each of its available subspecies. Similarly, *T. gracilis*, *T. minor*, *T. tana*, *T. nicobarica*, and others are represented by one specimen from each of their subspecies examined. In contrast, the two taxa *U. everetti*, a monotypic species, and *P. lowii*, represented in the original study sample by one subspecies, were represented in the analysis by two and three specimens, respectively. After examining all available specimens, the author felt that the range of morphological variation for each polymorphic species could be adequately represented by selecting one specimen for each form, with the exceptions noted above. While such a sample is adequate for between-group comparisons, it precludes the possibility of within-group comparisons. A more detailed analysis of within-group variation in tree shrews is presently in progress.

Figure 3 presents a simplified matrix of association coefficients for the specimens. The seven clusters of specimens identified (and separated in the figure by horizontal lines) have a high degree of within-group association (the within-group association coefficients exceeding 0.69) in contrast to the markedly-reduced degree of association between groups.

The right-hand portion of figure 4 presents a dendrogram of the 49 specimens by number (the numbers and corresponding names are given in fig. 3), with the seven clusters of figure 3 distinguished at the 0.76 level of association. The largest cluster contains all specimens (1–30) of *T. glis*, *T. gracilis*, *T. montana*, *T. mulleri,* and *T. splendidula*. While *T. montana* and *T. splendidula* are very similar dentally to most *T. glis*, *T. gracilis* and *T. glis versurae* (specimens 28, 29, 30) have lower levels of association with the rest of the group. The distinctiveness of *T. g. versurae* from the other forms of *T. glis* is rather surprising, but it is one of the smaller *T. glis* subspecies and tends to approach *T. gracilis* in overall body-size as well as in dentition.

Table II. List of the dental traits used in the Q-mode and R-mode analyses. The left column gives a short description of each trait, and the right column lists the code for character states of each trait

Trait	Character states
1. Relative length of upper incisors	0, I^1 and I^2 equal in length; 1, I^1 longer than I^2; 2, I^2 longer than I^1
2. Relative size of upper canine	0, equal to P^3; 1, greater than P^3
3. Relative size of P^2	0, less than P^3; 1, equal to P^3
4. Relative size of I_1 and I_2	0, I_1 and I_2 approximately equal in size; 1, I_2 significantly larger than I_1
5. Relative length of I_3	0, I_3 less than 1/2 length of I_2; 1, greater than 1/2 length of I_2
6. Relative size of P_2	0, smaller than P_3; 1, greater than P_3
7. Shape of I^2	0, simple cone; 1, premolariform; 2, caniniform
8. I^1 reflected mesially	0, no; 1, yes
9. Shape of upper canine	0, simple cone; 1, premolariform
10. Shape of cusps on M^1	0, with well-defined cristae; 1, more bunodont
11. Shape of lower canine	0, incisorform; 1, caniniform; 2, premolariform
12. Upper canine has two roots	0, no; 1, yes
13. Root number of P_2	0, one root; 1, two roots
14. Root number of P^3	0, two roots; 1, three roots
15. Root number of P_3	0, one root; 1, two roots
16. Protocone present on P^3	0, no; 1, yes
17. Parastyle present on P^3	0, no; 1, yes
18. Metastyle present on P^3	0, no; 1, yes
19. Preparacrista present on P^3	0, no; 1, yes
20. Postparacrista present on P^3	0, no; 1, yes
21. Stylar shelf present on P^4	0, no; 1, yes
22. Preparacrista present on P^4	0, no; 1, yes
23. Postparacrista present on P^4	0, no; 1, yes
24. Hypocone present on P^4	0, no; 1, yes
25. Hypocone present on M^1	0, no; 1, present but poorly defined; 2, well developed
26. Nature of mesostyle on M^1	0, single; 1, bifid
27. Internal cingulum present on M^1	0, no; 1, yes
28. Hypocone present on M^2	0, no; 1, present but poorly defined; 2, well developed
29. Nature of mesostyle on M^2	0, single; 1, bifid
30. Internal cingulum present on M^2	0, no; 1, yes
31. P^2 has a posterior fovea	0, no; 1, yes
32. P_3 has an anterior fovea	0, no; 1, yes

Table II (continued)

Trait	Character states
33. P_3 has a talonid shelf	0, no; 1, yes
34. P_4 has three cusps on trigonid	0, no; 1, yes
35. P_4 talonid divided	0, no; 1, yes
36. M_1 has an entoconulid	0, no; 1, yes
37. M_1 has a precingulum	0, no; 1, yes
38. M_1 has a cingulum	0, no; 1, yes
39. M_2 has an entoconulid	0, no; 1, yes
40. M_2 has a precingulum	0, no; 1, yes
41. M_2 has a cingulum	0, no; 1, yes
42. M_3 has a hypoconulid	0, no; 1, yes
43. M_3 has a precingulum	0, no; 1, yes

The second cluster (specimens 31–38) consists of the subspecies of *T. tana* and the two specimens of *U. everetti*. The third cluster consists of a single specimen (39), *T. picta*, which is distinct from all other specimens at the 0.76 level of association. The fourth cluster consists of the specimens of *T. minor* and *T. javanica* (40, 41, 42), the fifth cluster of the specimens (43, 44) of *Dendrogale*; and the specimens (45, 46) of *T. nicobarica* are associated in the sixth cluster. The three specimens (47, 48, 49) of *P. lowii* form the last (seventh) cluster and are distinct from the subfamily Tupaiinae at the 0.55 level of association.

Since the present study is confined to an analysis of the dentition of the tree shrews, the cluster analyses and the association coefficients were not expected to demonstrate exact taxonomic relationships. However, several of the relationships are certainly interesting and suggestive of areas for future taxonomic concern. Of particular note is the dissimilarity between the *T. tana* cluster and the *T. glis (et al.)* cluster in respect to their dentition. As mentioned earlier, the relationship of *T. tana* to the rest of the *Tupaia* has been quite problematical. On the basis of a comparison of their dentitions, the pronounced distinctiveness of *T. tana* is emphasized; but the similar degree of distinctiveness between *T. nicobarica* and *T. minor/T. javanica* suggests that, if a subgeneric separation of *T. tana* is accepted, then *T. nicobarica* and *T. minor/T. javanica* should also be separated at the same taxonomic level. However, these points need confirmation from other data before any taxonomic revisions are suggested.

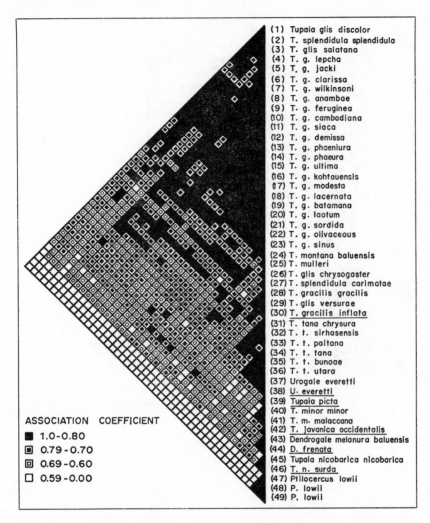

(1) Tupaia glis discolor
(2) T. splendidula splendidula
(3) T. glis salatana
(4) T. g. lepcha
(5) T. g. jacki
(6) T. g. clarissa
(7) T. g. wilkinsoni
(8) T. g. anambae
(9) T. g. feruginea
(10) T. g. cambodiana
(11) T. g. siaca
(12) T. g. demissa
(13) T. g. phaeniura
(14) T. g. phaeura
(15) T. g. ultima
(16) T. g. kohtauensis
(17) T. g. modesta
(18) T. g. lacernata
(19) T. g. batamana
(20) T. g. laotum
(21) T. g. sordida
(22) T. g. olivaceous
(23) T. g. sinus
(24) T. montana baluensis
(25) T. mulleri
(26) T. glis chrysogaster
(27) T. splendidula carimatae
(28) T. gracilis gracilis
(29) T. glis versurae
(30) T. gracilis inflata
(31) T. tana chrysura
(32) T. t. sirhasensis
(33) T. t. paitana
(34) T. t. tana
(35) T. t. bunoae
(36) T. t. utara
(37) Urogale everetti
(38) U. everetti
(39) Tupaia picta
(40) T. minor minor
(41) T. m. malaccana
(42) T. javanica occidentalis
(43) Dendrogale melanura baluensis
(44) D. frenata
(45) Tupaia nicobarica nicobarica
(46) T. n. surda
(47) Ptilocercus lowii
(48) P. lowii
(49) P. lowii

ASSOCIATION   COEFFICIENT

■ 1.0-0.80
◨ 0.79 - 0.70
◻ 0.69 - 0.60
☐ 0.59 - 0.00

Fig. 3. A simplified matrix of association coefficients. The association coefficient for any two specimens can be found at the junction of their lines.

Equally as interesting as the separation of *T. tana* from the *T. glis* cluster is the dental similarity of *T. tana* to *Urogale*. These two groups are also similar in their elongated rostra, terrestrial habits, and in their greater preference for meat [SORENSON and CONAWAY, 1964]. LYON [1913], aware of the similarities, suggested that they may have arisen from a common ancestral stock. More recently, SZALAY [1969], in remarking on the

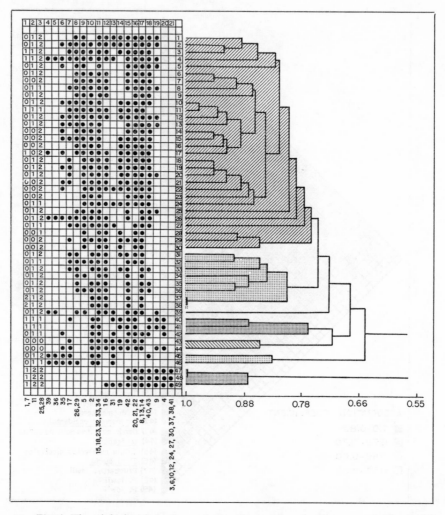

Fig. 4. The right-hand portion of the figure is a dendrogram of 49 tupaiid specimens with the clusters delimited at the 0.76 level of association. Clustered specimens 1–30 are *Tupaia glis, T. splendidula, T. montana, T. mulleri,* and *T. gracilis.* Specimens 31–36 are *T. tana;* 37 and 38 are *Urogale everetti;* 39 is *T. picta;* 40–42 are *T. minor* and *T. javanica;* 43 and 44 are *Dendrogale melanura* and *D. frenata;* 45 and 46 are *T. nicobarica;* and 47–49 are *Ptilocercus lowii.* (For a listing of all of the specimens and their corresponding numbers, refer to fig. 3.) The left half of the figure illustrates the distribution by specimen of the 43 traits used in the analyses; the traits are identified by number along the lower left edge of the figure. The lines indicating the 21 different patterns of distribution of the traits are numbered from left to right at the top of this part of the figure. A black dot indicates the presence

dietary preference of *T. tana* for meat, suggests that it is an example of a species in which behavioral differences preceded morphological differences. However, the morphological similarities of *Urogale* and *T. tana* noted here tend to refute SZALAY's contention that *T. tana* has adapted behaviorally but has yet to adapt morphologically.

The similarity of the dentitions of *T. minor* and *T. javanica* is also noteworthy, as is their shared dissimilarity from other species of *Tupaia*. Both species are also similar in size and in the possession of tails longer than their bodies. If MARTIN's [1968] observation that the more arboreal tree shrews, including *T. minor*, tend to have tails longer than their bodies is true, then one would expect *T. javanica* also to be an arboreal species. The dental similarity of the two species would add further support to the possibility that *T. javanica* is an arboreal species similar to *T. minor*, a suggestion that can only be confirmed by field observation.

The distinctiveness of the dentition of *T. nicobarica* is quite surprising and in many ways unexpected. However, LYON [1913] remarked on the aberrant morphology of this species, which is confined to the Nicobar Islands; and its dental distinctiveness provides additional data to support his statement. The taxonomic position of *T. nicobarica*, however, must await analysis until more work is done with the entire family.

Finally, it is necessary to consider the dissimilarity of the relationships suggested in the dendrogram from those indicated in the current classification of the tree shrews [NAPIER and NAPIER, 1967]. For example, while the distinctiveness of the dentition of at least four forms of *Tupaia* has been considered above, the inability to distinguish the species *T. gracilis*, *T. glis*, *T. montana*, and *T. splendidula* has not been discussed. Since the relationships presented here have been based on discrete traits of the dentition alone, the lack of separation of some species is to be expected, irrespective of the pleiotropic effect of some genes. Any dissimilarities among the species as indicated by size, for instance, have been considered only indirectly in connection with those dental traits that may be correlated with size; had size been taken into account more fully, there

---

of character state '1' for the trait in a particular specimen; a character state of '0' for the trait is indicated by a blank box for the corresponding specimen. The 5 traits listed at the far left in lines 1–3 are those with 3 character states, and the numbers in the boxes indicate the state present in a particular specimen. (A detailed listing of the traits with their corresponding numbers and an explanation of the coding of their possible character states are presented in table II.)

would have been little difficulty distinguishing the small *T. gracilis*. On the other hand, separation of *T. glis*, *T. montana*, and *T. splendidula* would still be difficult had size been the only additional characteristic considered.

B. Trait Analysis

In addition to the Q-mode analysis establishing the relationships of the specimens, the SMC technique was used for an R-mode analysis to determine the interrelationship of the traits. The purpose of the R-mode analysis was to identify the traits defining the various groups of specimens and to determine at what level of taxonomic analysis the various traits could be used in future identification of tupaiids.

The left side of figure 4 presents a summary of these data. The traits, listed in 21 lines along the lower left-hand edge of the figure, are identified by the numbers given them in table II. All traits listed together on one line showed the same distribution among the specimens. Thus, 21 different patterns of distribution of the 43 traits are apparent. The 21 lines or patterns of distribution are numbered from left to right at the top left edge of the figure. As already noted, the 49 tupaiid specimens are numbered consecutively from top to bottom in the center of the figure; the numbers correspond with the names listed in figure 3. A code of '1' for a trait in a particular specimen is indicated by a dot in the corresponding box. For instance, the ten traits listed together on line 21 (3, 6, 10, 12, 24, 27, 30, 37, 38, 41) are present in the three specimens (47, 48, 49) of *P. lowii* and absent in all other specimens examined. The traits listed in the first three lines, however, are those with three character states; and the code for the character state recorded for each specimen is shown in the corresponding box. Three types of distribution patterns of the traits become apparent when the data are analyzed in the above fashion. Traits listed in line 21 are restricted to the Ptilocercinae. Traits listed in lines 12–20 are found in some specimens throughout the family Tupaiidae, and traits listed in lines 4–11 are restricted to the subfamily Tupaiinae.

Of the 43 traits selected by the author, 20 (lines 8, 9, 10, 11, 21) primarily define the two subfamilies of tree shrews. Eight of these 20 (10, 26, 27, 29, 30, 37, 38, 41) define differences in the molars. Since only five additional two-state traits defining molar morphology (all of which concern the lower molars) seem to vary significantly below the level of the subfamily, one apparent conclusion is that it is extremely difficult to distinguish taxa below that level on the basis of discrete traits of the molars. VAN VALEN [1965] discussed this problem earlier, emphasizing the difficulty of identifying molars of tupaiids even at the level of the order.

The other 12 traits distinguishing the two subfamilies primarily concern the premolar series. Although minor variations in each premolar are effective in helping to distinguish the various Tupaiinae, more apparent differences, such as the molarization of the fourth upper premolar and the premolarization of the upper canine distinguish the subfamilies. The impression, then, is again that of two groups of animals with fairly distinctive dental adaptations. In *Ptilocercus*, the overall occlusal surface of the molars is increased by the distinctively larger hypocone. In addition, the fourth premolar, with the hypocone, is more molariform in this animal than it is in the Tupaiinae; and there is also the tendency for the inclusion in *Ptilocercus* of the upper canine in the premolar series.

A similar series of traits effectively differentiating the genera was not apparent; clear separation of the genera within the subfamily Tupaiinae was, therefore, unattainable. However, it is apparent from the trait analysis that variations in dental morphology among the subfamily Tupaiinae occur primarily in the anterior dentition, the incisors, canines, and premolars. The author suspects that a metrical analysis of palatal dimensions would also show much greater size and shape variations in the anterior part of the palate and in the face. If this suspicion is confirmed, it would be interesting to determine the adaptive and functional values of these differences.

The problem of determining traits useful for taxonomic differentiation at various levels of generalization has been one point of controversy for many years. While many phylogenetic taxonomists have felt the identification of such traits rests upon the determination of their biological importance and significance, others have felt that this approach was too subjective. Recently, FARRIS [1966, 1970] has reported a technique for determining such useful traits on the basis of their constancy or of the amount of low within-population variation. While his technique is effective for non-meristic traits (measurable traits such as palatal length, etc.), he points out that the effectiveness for discrete traits is limited to those having a sample variance greater than zero. Where possible, the discrete traits used in the present study were checked before statistical analysis to verify the low within-group variation for each population considered; and those with a high within-group variation at the population level were not included in the statistical analysis. However, at the subfamily level of generalization, a series of traits with low within-group variation has been identified for each subfamily (particularly those in lines 10, 11, and 21 of fig. 4) and are suggested for use in future studies of tree shrews at the same level of generalization.

The trait analysis can also be used to determine the traits or clusters of traits defining the various tree shrews. As noted above, only *Ptilocercus* has character state '1' for each of traits 3, 6, 10, 12, 24, 27, 30, 37, 38, and 41. Conversely, most members of the subfamily Tupaiinae can be identified by character state '1' for traits 2, 5, 15, 18, 23, 26, 29, 32, 33, and 34 (lines 8, 9, 10, 11). The other traits considered in the analysis are not of particular value in distinguishing the two subfamilies, for they define only portions of either subfamily or are generally characteristic of some members of both subfamilies.

Within the subfamily Tupaiinae some of the species can also be distinguished. *T. nicobarica* for instance is generally distinguished from the rest of the Tupaiinae by the presence of traits 17, 31, 36, and 39 (lines 4, 5, 7, 13). Although each of these traits are shared with some other members of the Tupaiinae, only *T. nicobarica* has all of them. Among other traits shared with some of the other Tupaiidae, *Dendrogale* is particularly characterized by traits 17, 19, and 31 (lines 7, 13, and 14). In addition, *Dendrogale* is unique in lacking hypocones on the upper first and second molars (traits 25 and 28, line 3). Among other traits, *T. minor* and *T. javanica* are particularly characterized by 9, 40, and 43 (lines 18, 19). Separation of *T. tana* and *Urogale* from the largest cluster, *T. glis (et al.)*, is not as marked. Primarily, the *T. tana* and *Urogale* dentitions are distinguished by traits 26 and 29 (line 8); while the dentitions of most members of the *T. glis* cluster are distinguished by traits 5, 17, 19, 40, and 43 (lines 7, 9, 14, 18). Within the *T. glis* cluster, *T. gracilis* and *T. g. versurae* can be distinguished generally by their lack of traits 40 and 43 (line 18).

Summary

The present paper discusses dental variability among the Tupaiidae. The first section of the report includes a qualitative analysis of the dental morphology of the tupaiids and a discussion of the range of variation of various traits. The second section discusses the results of a Q-mode cluster analysis of the tree shrews based on 43 discrete traits of the dentition. The resulting clusters, while not reflecting accurately the currently accepted classification of the Tupaiidae, differ principally in those areas which have proved controversial in the past. Specifically, the clusters indicate that the dentitions of *T. tana, T. minor, T. javanica,* and *T. nicobarica* differ markedly from the dentitions of *T. glis, T. gracilis, T. montana, T. mulleri,* and *T. splendidula,* with the dentition of the latter cluster showing a high level of within-group variation. The cluster analysis also indicates a close similarity in the dentitions of *T. tana* and *U. everetti.* In a similar fashion, *T. minor* and *T. javanica* are shown to have similar dentitions. The dentition of *Ptilocercus* is shown to be markedly distinct from the dentition of the Tupaiinae. The third section of the report discusses an R-mode analysis of the 43 dental traits; it emphasizes the interrelationships of the traits, identifies those characteristic of each group, and discusses at what levels of analysis the traits are effective in discriminating the tree shrews.

Acknowledgements

The present research was made possible by a Smithsonian Institution post-doctorate fellowship administered through the Department of Anthropology, U.S. National Museum. The author would like to extend special thanks to J. L. ANGEL, Department of Anthropology, and R. W. THORINGTON, Director of the Primate Biology Program, for making available to the author the excellent facilities and collections of the Museum. Special acknowledgements also go to R. MACPHEE, T. NICKS, D. J. ORTNER, K. M. STEELE, R. W. THORINGTON, and D. UBELACKER for many helpful discussions pertinent to the research and presentation of the report. Generous help with the statistical aspects of the research was given by H. DANIEL ROTH, Information Systems Division of the Smithsonian Institution, and R. CORRUCCINNI, University of California, Berkeley.

References

BOHLIN, B.: Some mammalian remains from Shih-erh-ma-ch'eng, Hui-hui-p'u area, Western Kansu. Report of the sino-Swedish scientific expedition to the northwestern provinces of China, vol. 35, pp. 1–47 (1951).

CAMPBELL, C. B. G.: Pyramidal tracts in primate taxonomy; Diss., Univ. of Illinois, Urbana (1965).

CAMPBELL, C. B. G.: The relationships of the nervous system. Evolution *20:* 276–287 (1966).

CARLSSON, A.: Über die Tupaiidae und ihre Beziehungen zu den Insektivora und den Prosimiae. Acta zool. *3:* 264 (1922).

CHASEN, F. N.: A handlist of Malaysian mammals. Bull. Raffles Mus. *15:* 1–209 (1940).

CHASEN, F. N. and KLOSS, C. B.: On a collection of mammals from the lowlands and islands of North Borneo. Bull. Raffles Mus. *6:* 1–82 (1931).

Conisbee, L. R.: A list of the names proposed for genera and subgenera of recent mammals (Brit. Mus. natur. Hist., London 1953).

Davis, D. D.: Notes on the anatomy of the tree shrew *Dendrogale*. Field Mus. natur. Hist. Publ. Zool. *20:* 383–407 (1938).

Davis, D. D.: Mammals of the lowland rain-forest of North Borneo. Bull. nat. Mus., Singapore *31:* 1–129 (1962).

Ellerman, J. R. and Morrison-Scott, T. C. S.: Checklist of Paleartic and Indian mammals (Brit. Mus. natur. Hist., London 1951).

Elliot, O. S.: The biology of tree shrews with an emphasis on *Tupaia glis* (Diard 1820) of Malaya; Diss., Harvard Univ. Cambridge (1969).

Farris, J. S.: Estimation of conservatism of characters by constancy within biological populations. Evolution *20:* 587–591 (1966).

Farris, J. S.: On the relationship between variation and conservatism. Evolution *24:* 825–827 (1970).

Fiedler, W.: Übersicht über das System der Primaten; in Hofer, Schultz und Starck Primatologia, vol. 1, pp. 1–266 (Karger, Basel 1956).

Gregory, W. K.: The orders of mammals. Bull. amer. Mus. natur. Hist. *27:* 1–524 (1910).

Hiiemäe, K. M. and Kay, R. F.: Evolutionary trends in the dynamics of primate mastication; in Zingeser Craniofacial biology of primates. Symp. Proc. 4th int. Congr. Primat., Portland, Oreg. 1972, vol. 3, pp. 28–64 (Karger, Basel 1973).

Hill, J. E.: The Robinson collection of Malaysian mammals. Bull. Raffles Mus. *29:* 1–112 (1960).

Hill, J. P.: On the placentation of *Tupaia*. J. Zool., Lond. *146:* 278–304 (1965).

Kindahl, M.: On the development of teeth in *Tupaia javanica*. Ark. Zool. *10:* 463–479 (1957).

Kloss, C. B.: On a fourth collection of Siamese mammals. J. natur. Hist. Soc. Siam *3:* 49–69 (1918).

Le Gros Clark, W. E.: On the brain of *Tupaia minor*. Proc. zool. Soc., Lond. *44:* 1053–1074 (1924).

Le Gros Clark, W. E.: On the skull of *Tupaia*. Proc. zool. Soc., Lond. *45:* 559–567 (1925).

Luckett, W. P.: Placentation of the tree shrews (family Tupaiidae). Anat. Rec. *157:* 280 (1967).

Lyon, M. W.: Tree shrews: an account of the mammalian family Tupaiidae. Proc. U.S. nat. Mus. *45:* 1–188 (1913).

Martin, R. D.: Tree shrews: unique reproductive mechanism of systematic importance. Science *152:* 1402–1404 (1966).

Martin, R. D.: Reproduction and ontogeny in tree shrews *(Tupaia belangeri)*, with reference to their general behavior and taxonomic relationships. Z. Tierpsychol. *5:* 409–532 (1968).

Mayr, E.: Animal species and evolution (Belknap Press of Harvard Univ. Press, Cambridge 1963).

McKenna, M. C.: The early Tertiary mammals and their ancestors; in Moore 16th int. Congr. Zool., vol. 4, pp. 69–74 (Washington, D.C. 1963).

McKenna, M. C.: Paleontology and the origin of the primates. Folia primat. *4:* 1–25 (1966).

MEDWAY, LORD G. G. H.: The status of *Tupaia splendidula* Gray, 1865. Treubia *25:* 269–272 (1961).

MEDWAY, LORD G. G. H.: Mammals of Borneo: field keys and annotated checklist (Malaysian Branch, roy. asiatic Soc., Singapore 1965).

MEISTER, W. and DAVIS, D. D.: Placentation of the pygmy tree shrew, *Tupaia minor*. Fialdiana, Zool. *35:* 73–84 (1956).

MIVART, ST. G.: Note on the osteology of Insectivora. Part 1. J. Anat. Physiol. *1:* 281–312 (1867).

NAPIER, J. R. and NAPIER, P. H.: A handbook of living primates (Academic Press, London 1967).

SABAN, R.: Les affinités du genre *Tupaia* Raffles 1821, d'après les caractères morphologiques de la tête osseuse. Ann. Paléont. *42:* 169–224 (1956); *43:* 1–44 (1957).

SIMPSON, G. G.: A new insectivore from the Oligocene, Ulun Gochu horizon of Mongolia. Amer. Mus. Novitates *505:* 1–22 (1931).

SIMPSON, G. G.: The Tiffiany fauna, Upper Paleocene. II. Structure and relationships of *Plesiadapis*. Amer. Mus. Novitates *816:* 1–30 (1935).

SIMPSON, G. G.: The principles of classification and a classification of mammals. Bull. amer. Mus. natur. Hist. *85:* 1–350 (1945).

SOKAL, R. R. and MICHENER, C. D.: A statistical method for evaluating systematic relationships. Univ. Kansas Sci. Bull. No. 38, pp. 1409–1438 (1958).

SOKAL, R. R. and SNEATH, P. H.: Principles of numerical taxonomy. (Freeman, San Francisco 1963).

SORENSON, M. W. and CONAWAY, C. H.: Observation of tree shrews in captivity. Sabah Soc. J. *2:* 77–91 (1964).

SORENSON, M. W. and CONAWAY, C. H.: Observations on the social behavior of tree shrews in captivity. Folia primat. *4:* 124–145 (1966).

SZALAY, F. S.: The beginning of primates. Evolution *22:* 19–36 (1968).

SZALEY, F. S.: The Mixodectidae, Microsyopidae and the insectivore-primate transition. Bull. amer. Mus. natur. Hist. *140:* 195–330 (1969).

VANDENBERGH, J. G.: Feeding, activity and social behavior of the tree shrews, *Tupaia glis*, in a large outdoor enclosure. Folia primat. *1:* 199–207 (1963).

VAN VALEN, L.: Tree shrews, primates and fossils. Evolution *19:* 137–151 (1965).

VAN VALEN, L.: Deltatheridia, a new order of mammals. Bull. amer. Mus. natur. Hist. *132:* 1–126 (1966).

WEISHART, D.: Fortran II programs for eight methods of cluster analysis (Clustani). Computer Contribution 38 (State Geol. Survey, University of Kansas, Lawrence 1969).

Author's address: Dr. D. GENTRY STEELE, Department of Anthropology, University of Alberta, *Edmonton 7* (Canada)

Craniofacial Growth, Development, and Functional Anatomy

Symp. IVth Int. Congr. Primat., vol. 3: Craniofacial Biology of Primates, pp. 180–190 (Karger, Basel 1973)

Remodeling of the Craniofacial Skeleton Produced by Orthodontic Forces

B. MOFFETT

Department of Orthodontics, University of Washington, Seattle

Research on nonhuman primates is a continuing intellectual contest between the primate investigator and the primate subject, with success going to whichever individual is more adaptable. For the orthodontic investigator this creates some problems similar to those experienced with human patients. On the other hand, it also provides a source of information on craniofacial growth and remodeling that is most applicable to the solving of clinical problems and cannot be obtained elsewhere. These themes can be illustrated in the following summary of studies carried out largely by orthodontic graduate students at the University of Washington.[1]

Research Methodology

A technical problem which soon becomes evident in orthodontic studies on monkeys results from the animals' tendency to eliminate the orthodontic appliance by any means available – either by removing it with hands or feet, by scraping it against the cage, or by destroying it with

1   These research projects are listed at the end of the article, but are not cited individually because my comments represent an integration of the separate studies. Each investigation listed is an MSD. thesis in Orthodontics. Most were made possible by means of the Affiliate Investigator program at the Regional Primate Research Center, Seattle, Wash. They were supported in part by Public Health Service grants HD-02280 and RR-00166, the Poncin Scholarship Fund, and the University of Washington Orthodontic Memorial Fund.

occlusal forces, and so on. How much of a problem this will be depends to a large extent on the behavioral pattern or personality of the monkey, which varies greatly from one animal to another within the same species. The animal handler can often advise the investigator as to which specific monkey has favorable or unfavorable personality traits for a particular investigation. In any case, the best policy is to provide for daily surveillance and checking of the appliance during the entire experimental period and, if necessary, to use an appropriate form of physical restraint. The latter can be accomplished in various ways depending upon the type of appliance and the species, size, and personality of the animal. Such restraint ranges from none (except that required for animal care) to the use of a 'chair' in which the monkey is free to sit or stand but cannot reach his head. Lesser degrees of restraint can be provided by the use of 'boxing glove' bandages on hands and feet (this technique has been used successfully on squirrel monkeys) or by the wearing of a chest harness on which a plate-like collar is mounted to prevent the animal from reaching his head (used successfully on rhesus monkeys).

Our purpose in designing orthodontic experiments in monkeys has been to alter the morphology of the craniofacial skeleton by means of mechanical forces applied to the jaws of both growing and adult animals. What we are studying, therefore, is a biologic response to a mechanical stimulus; and in order for the findings to have clinical application, the results must be documented at both the histologic and morphologic levels.

Histologic documentation identifies the cellular mechanisms responsible for morphologic alteration; but it does not provide satisfactory evidence, *per se*, that a morphologic alteration of clinical significance did, in fact, occur. In order for experimentally-produced craniofacial remodeling to have an established clinical significance, the resulting morphologic alteration must reach the gross or macroscopic level and naturally must be greater than the error of measurement.

There are two methods for obtaining this type of morphologic documentation in monkeys, each with its advantages and disadvantages. The procedure most commonly used is cephalometric roentgenography[2], and its main advantage is that serial changes can be documented in the same

2   Cephalometric roentgenography is a standardized X-ray technique for obtaining reproducible radiographs of the skull. From such a radiograph a penciled tracing is made of significant bony contours and reference planes. This tracing is used for measurements of head size and shape and for comparison with other films taken of the same animal at different time periods.

animal during the course of an experiment as well as at its termination. However, in order to apply this technique effectively to monkeys, modifications must be made in the procedure conventionally used for clinical or human cephalometrics.

Modifications are necessary because the error in head positioning by means of ear rods is as large, or even larger, with monkeys as it is for humans, even though the monkey cranium is much smaller. The net result is that the measurement error in cephalometric tracings of monkeys will often be as large as the experimentally produced morphologic change. In order to recognize this positioning error when it is present, metallic implant markers should be placed in the maxilla and mandible of the monkey. These markers will also allow one to superimpose serial tracings in a way which can differentiate between relocation and remodeling of the implanted bones. For the latter reason alone, such implants must be used in studies of orthodontic remodeling.

In order to eliminate the positioning error, we now use a small coupling device surgically attached to the frontal bone as a substitute for the ear rods. By this technique the monkey can be positioned repeatedly in the headholder in exactly the same orientation. The way in which the investigator may wish to solve this problem is not critical, but he must document the morphologic alterations with a precision greater than that considered clinically acceptable for human beings.

The other method of morphologic documentation which can be used is the preparation of a dry bone specimen or macerated skull. With this technique only final and not serial measurements can be made; however, the experimentally-produced changes can be studied in three dimensions. Such study will frequently reveal an experimental alteration that would be difficult or even impossible to detect in radiographs and histologic sections.

Although priorities of content restrict further discussion of this topic, much more can and should be said in regard to the need and technique for cephalometric documentation in experimental studies of skeletal remodeling. Once the investigator has become familiar with the associated histologic mechanisms, he will learn more about clinical applicability of skeletal remodeling from the morphologic analysis than can be gained from microscopic study. My recommendation would be that histologic specimens be prepared from areas where the morphologic analysis has produced findings of critical importance. The histologic data should be studied to see whether they are compatible with the morphologic analysis, provided the cellular mechanisms for such changes are already understood.

Because the techniques and productivity of *in vivo* bone labeling are more widely known, this aspect of skeletal analysis will not be discussed except to say that it should be used routinely in all studies of bone growth and remodeling. From such a histologic documentation of bone deposition, and indirectly of bone resorption, experimental studies in monkeys will answer the orthodontist's basic questions: How long must I use this orthodontic appliance in order to produce a skeletal response? How long will this skeletal response continue after removal of the orthodontic appliance? This fundamental aspect of orthodontic treatment planning still remains unanswered.

Clinical Applicability

Why use nonhuman primates for craniofacial research? They are well known as expensive, uncooperative animals, highly susceptible to human infections (and vice versa); and undoubtedly many basic questions in craniofacial biology still remain that could be adequately answered with nonprimate mammals. For instance, the cellular mechanisms and biologic signals responsible for mechanically induced skeletal remodeling are undoubtedly identical for the majority of, if not all, mammals and need only be confirmed at the primate level.

Regardless of these truths, for many research questions involving skeletal biology and morphology, a nonhuman primate provides the best experimental model from which the findings can be transferred to man. Consider the question of cleft palate teratology, which has been so thoroughly explored in laboratory rodents. When the numerous teratogens producing palatal clefts in rats and mice are tested on carefully-timed monkey embryos, the result so far has been mainly either no effect or, at higher doses, spontaneous abortion. Continued studies on primate teratology will reveal and help to explain a protective mechanism against maldevelopment that is nonexistent or rudimentary in nonprimate mammals.

The development and testing of new orthodontic and surgical techniques for dealing with occlusal problems in man are most easily accomplished in monkeys. The following procedures were all worked out on monkeys before we employed them in the treatment of human patients: the use of metallic implant markers for documentation and analysis of

growth patterns and treatment results; the surgical interruption of trans-
septal fibers to prevent postorthodontic rotational relapse; and the remodeling of facial sutures by means of extraoral traction at high force levels.

Because orthodontic studies often represent problems in bioengineer-
ing, the morphologic and physiologic similarities of the simian and human
craniofacial complex provide an important advantage over nonprimate
models. However, even the slight developmental and morphologic differences which are superimposed on the basic similarities between monkeys
and man represent valuable research opportunities. For example, the failure of malocclusions to occur spontaneously in monkeys is a potential
natural experiment that can be manipulated by the investigator to help
explain malocclusion in man. It seems likely that the less variable occlusal relationship in monkeys results more from the protective influence of
dental intercuspation than from a genetic mechanism controlling the size
of the upper and lower jaws. In this instance, moreover, a study of occlusion in monkeys seem more meaningful for application to man than does
information transferred from genetic analyses in dogs and other nonpri-
mates.

Several other advantages of primate research models deserve mention. There is a great range in size from one species of monkey to another;
the tiny squirrel monkey, for example, which as an adult weighs 600 g
and is thus small enough for labeling with radioisotopes but still large
enough for orthodontic expansion of the palatal suture; or the 15-kg adult
macaque with teeth and cranium of sufficient size to allow use of all types
of fixed and removable intraoral and extraoral appliances. In addition to
their generally smaller size, the time required for growth in various nonhuman primates shows an advantageous range of shortening compared to
human growth. The growth period in squirrel monkeys, though not yet
documented, appears to last only one or two years; in the rhesus it is
completed in approximately six years. In each case the growth period
can be subdivided into stages of deciduous, mixed, and permanent dentitions, just as in man; and experimental alterations of craniofacial growth
can be produced during any of these developmental stages.

Present Findings

The craniofacial skeleton grows and functions in an environment of
mechanical forces that influences the shape and relative position of each

bone in the complex. A shifting equilibrium is maintained between this physical environment and the morphology of the skeleton through the process of biologic adaptation called skeletal remodeling. The forces that stimulate this adaptive response originate in many ways; and they vary in magnitude, duration, and direction. Included among them are muscle activity, resistance of adjacent tissues, gravity, forces generated by manmade appliances, and others. In order to study the nature of this adaptive response, the investigator must alter the existing equilibrium in a way that can be documented. For this reason we have chosen to study the effects of manmade applicances that allow us to produce forces of known direction, magnitude, and duration.

When the positions of bones and teeth within the craniofacial complex are forcibly altered with orthodontic appliances, the forces so generated tend to be concentrated at the site of force application and at the articular surfaces of the displaced components. It is not surprising, therefore, that the resulting biologic adaptation is greatest at these surfaces. The major exception to this principle is the fact that when force is applied directly to the crowns of erupted teeth, no surface remodeling occurs at the point of force application because the dental crowns are no longer covered by cells capable of depositing or resorbing enamel. In this instance, the biologic remodeling associated with orthodontic tooth movement occurs mainly at the articular surfaces between the dental root and the surrounding alveolar bone.

Usually, when mechanically-produced remodeling activity is seen on an articular surface, the adjacent nonarticular surfaces also may show morphologic changes. However, these changes are smaller and generally diminish rapidly with distance from the joint. The magnitude of articular remodeling and, thus, of the adjacent nonarticular surface depends on the type of articular tissue present in each joint. These circumstances make it necessary, therefore, to abandon the conventional descriptive approach used in osteology and to analyze the craniofacial skeleton as a complex of various types of articulations separated by bones. Only through this approach can one understand the nature of orthodontically-induced skeletal remodeling as it is practiced today.

For our purpose, the articulations of the skull can be classified into three anatomical categories based mainly on the type of articular tissue present: fibrous, cartilaginous, and synovial. The two examples of fibrous joints important to orthodontics are periodontal joints and sutures; the former connect the dental roots to the surrounding alveolar bone, the latter join together the individual facial and cranial bones. In both types of fibrous joints the skeletal components are connected by a well-vascularized collagenous ligament. This type of articular tissue is very responsive to mechanical forces: sustained stretching stimulates bone formation by mineralization of the collagen where the ligament inserts into bone; sustained compression of the ligament produced by forcing the two skeletal components together results in resorption of bone and to a lesser extent cementum on the dental root. The ease with which the periodontal joint can be mechanically remodeled forms the basis of orthodontic treatment. However, experiments on monkeys have shown that the sutural joints are just as easily remodeled; any orthodontic appliance which transmits force beyond the periodontal joints will produce similar remodeling activity in the facial and cranial sutures. This principle finds frequent orthodontic application when the midpalatal suture is mechanically stretched in order to increase the width of the upper dental arch. Furthermore, when extraoral appliances are used that displace the upper jaw in a certain direction, the position of the maxilla becomes changed by means of sutural remodeling.

The cartilaginous joints of orthodontic importance are the cranial base synchondroses and the cartilaginous growth center of the mandibular condyle. As the classification indicates, they consist of a junction between bone and cartilage. Both of these examples are temporary joints that last only while that part of the skeleton is growing by means of endochondral ossification, a process in which the cartilage is slowly mineralized, resorbed, and replaced by bone. Cartilage is basically an avascular tissue in which all metabolic activity is supported by the slow diffusion of fluids through the intercellular matrix. The matrix of cartilage contains a protein and carbohydrate complex, chondroitin sulfate, that resists compression very effectively and actually shows an increase in concentration when exposed to articular compression. For these reasons, cartilaginous joints are much more difficult to remodel compared to fibrous joints. In experiments done so far, the rate of growth occurring in either a cranial synchondrosis or the mandibular condyle has not been conclusively increased or decreased by means of mechanical force. However, sufficient experimental and clinical evidence exists to show that the direction of cartilaginous growth in these joints can be altered by orthodontic forces.

In the category of synovial joints, only one example is dealt with orthodontically, the temporomandibular joint. Synovial joints are so named because their skeletal components are joined together by a sleeve of collagenous tissue, the interior of which is partially lined with a lubricating tissue and fluid called synovium. The articular ends of the bones in such a joint are covered with a layer of cartilage or a similar nonvascular tissue, but this tissue does not extend uninterrupted across the joint as in the fibrous and cartilaginous joints. A lubricated articular cavity intervenes which gives the synovial joint a large range of motion. Furthermore, the avascular articular tissue provides a relatively stable articular contour even when the joint is exposed to mechanical forces of external origin. Our experimental results in causing the simian temporomandibular joint (TMJ) to remodel itself can be summarized as follows:

(1) The articular surface of the TMJ has shown no remodeling responses to orthodontic force in young growing monkeys. However, compressive force applied to the TMJ of adult monkeys has produced histologic evidence of remodeling at the area of maximal articular compression. This age difference suggests that the adult joint shows a decreased stability under mechanical loading and may in fact be susceptible to mechanically-induced arthritic changes. This condition may be due to the absence of the underlying condylar growth cartilage in the adult and to age change in connective tissue metabolism.

(2) Orthodontic posterior displacement of the mandible in growing and adult monkeys produced bone resorption where the posterior surface of the condyle was compressed against the postglenoid tubercle.

(3) Orthodontic anterior displacement of the mandible in growing monkeys produced histologic evidence that condylar growth had been directed more posteriorly.

The greatest significance of the findings reviewed above is that the methodology and techniques for orthodontic research in primates have been developed and proven effective, that a number of these findings have already been found applicable to the treatment of human patients, and that potential solutions have appeared for present problems involving growth and development of the craniofacial skeleton.

Future Goals

The development of therapeutic techniques will continue to be the most obvious goal of orthodontic research on primates. The broadest horizon visible in this area will be applying the principles learned in articular remodeling to periosteal surfaces. We have already demonstrated experimentally in monkeys that periosteal bone formation can be stimulated by applying a tensile or lifting force directly to periosteum. The biologic mechanism is identical to that seen in orthodontic remodeling of the periodontal and sutural joints; and I am convinced that techniques based on therapeutic periosteal remodeling will solve some of the major problems in orthodontics, prosthodontics, and periodontics.

On the biologic horizon numerous goals can be identified for primate research. There is still no general agreement on whether the human TMJ is loaded or subjected to articular compression during masticatory function. This controversy will continue until functional intra-articular pressure is actually measured by means of transducers placed in the primate TMJ. Several still unexplained aspects of craniofacial growth, such as the physiologic migration of muscle attachments and of teeth (eruption, mesial drift), should be analyzed by means of primate studies designed to show evidence of periosteal movement relative to the underlying bony surface. Already mentioned is the need in teratology to study the additional protective mechanisms that seem to be acting against maldevelopment in primates.

While directing our thoughts to the future, we should consider an aspect of skeletal remodeling which lends itself to nonhuman primate research: the effect of behavior and, therefore, of behavioral modification on the development and morphology of the craniofacial skeleton. The techniques in behavioral psychology and physiology which are now being used to elicit documented levels of performance in monkeys under controlled and reproducible conditions provide an ideal tool for studying the relationships between form and function. In most experimental studies on morphology, the influence of behavior and the fact that it may have been significantly altered during the experiment is either ignored completely or considered unimportant. It seems likely, however, that many of our present experimental results may be clarified and more correctly interpreted if we monitor behavior in all studies in which a morphologic alteration has been produced and if we design some studies in which altered behavior is the only experimental variable. The immediate applicability of this concept to clinical problems in craniofacial growth identifies nonhuman primates as the best research model available.

We hope that this review of methodology, findings, and future goals will recruit additional investigators into orthodontic research on nonhuman primates. It is a field in which the resources and opportunities available have developed faster than has the number of biologically-oriented researchers who are applying an adequate methodology to these problems.

References

ADAMS, C.: The effects of continuous posterior mandibular forces (class III) on the temporomandibular joint and the dentofacial skeleton of the *Macaca mulatta;* MSD thesis, University of Washington, Seattle (1969).

BOESE, L.: Increased stability of orthodontically rotated teeth following gingivectomy; MSD thesis, University of Washington, Seattle (1968).

COLLINS, A.: Influence of the gingival fiber complex on the relapse of orthodontically repositioned teeth in young rhesus monkeys; MSD thesis, University of Washington, Seattle (1969).

CUTLER, B.: Dentofacial changes produced by a modified Milwaukee brace in *Macaca mulatta:* roentgenographic and histologic studies; MSD thesis, University of Washington, Seattle (1968).

DART, J.: A histologic study of experimental intrusion of multirooted teeth in the *Macaca mulatta* monkey; MSD thesis, University of Washington, Seattle (1966).

ERICKSON, L.: Facial growth in the macaque monkey: a longitudinal cephalometric roentgenographic study using metallic implants; MSD thesis, University of Washington, Seattle (1958).

FERGUSON, R.: Influence of external forces on the positional interrelationship of the temporal bone and craniofacial sutures; MSD thesis, University of Washington, Seattle (1972).

FREDRICK, D.: Dentofacial changes produced by extraoral highpull traction to the maxilla of the *Macaca mulatta:* a histologic and serial cephalometric study; MSD thesis, University of Washington, Seattle (1969).

HANSEL, J.: A cephalometric and histologic evaluation of a new headholding device for serial cephalometric roentgenology on the rhesus *(Macaca mulatta)* monkey; MSD thesis, University of Washington, Seattle (1970).

HASSIG, F.: A cephalometric and histologic study in *Macaca mulatta* of the changes in the craniofacial complex during and following modified Milwaukee brace therapy; MSD thesis, University of Washington, Seattle (1969).

HENDERSON, P.: A histologic study of bone growth and remodeling in the maxilla on the *Macaca mulatta* monkey; MSD thesis, University of Washington, Seattle (1967).

JOHO, J.: The effects of extraoral lowpull traction to the mandibular dentition of *Macaca mulatta;* MSD thesis, University of Washington, Seattle (1971).

MEIKLE, M.: The effect of a class II intermaxillary force upon the dentofacial complex in the adult *Macaca mulatta* monkey; MSD thesis, University of Washington, Seattle (1969).

MERRILL, O.: The calcification pattern of the developing permanent dentition of the

Macaca nemestrina monkey as related to chronological age; MSD thesis, University of Washington, Seattle (1968).

MOORE, G.: A longitudinal study of thumb-sucking and open-bite in the *Macaca mulatta;* MSD thesis, University of Washington, Seattle (1970).

NORWICK, K.: The effect of reciprocal intermaxillary forces (class II) on the growing dentofacial complex in the *Macaca mulatta;* MSD thesis, University of Washington, Seattle (1969).

PIHL, E.: A serial study of the growth of various cranial and facial bones in the macaque monkey; MSD thesis, University of Washington, Seattle (1959).

RENSCH, J.: Direct cementation of orthodontic attachments; MSD thesis, University of Washington, Seattle (1972).

SENDROY, P.: A study of controlled tooth intrusion in the *Macaca nemestrina* monkey; MSD thesis, University of Washington, Seattle (1968).

SILVA, A.: A histologic study of normal facial growth and remodeling in the squirrel monkey *(Saimiri sciureus);* MSD thesis, University of Washington, Seattle (1968).

SPROULE, W.: Dentofacial changes produced by extraoral cervical traction to the maxilla of the *Macaca mulatta,* a histologic and serial cephalometric study; MSD thesis, University of Washington, Seattle (1968).

THOM, T.: A roentgenographic cephalometric study of craniofacial variability in the squirrel monkey *(Saimiri sciureus);* MSD thesis, University of Washington, Seattle (1965).

TURPIN, D.: Growth and remodeling of the mandible in the *Macaca mulatta* monkey; MSD thesis, University of Washington, Seattle (1966).

ZIMMERMANN, H.: The normal growth and remodeling of the temporomandibular joint of *Macaca mulatta;* MSD thesis, University of Washington, Seattle (1971).

Author's address: Dr. BEN C. MOFFETT, Department of Orthodontics, University of Washington, School of Dentistry, *Seattle, WA 98195* (USA)

Symp. IVth Int. Congr. Primat., vol. 3: Craniofacial Biology of Primates,
pp. 191–208 (Karger, Basel 1973)

A Functional Cranial Analysis of Primate Craniofacial Growth

M. L. Moss[1]

Department of Anatomy, College of Physicians and Surgeons and School of
Dental and Oral Surgery, Columbia University, New York

Introduction

The method of functional cranial analysis, as developed in this laboratory during two decades [Moss, 1971b, 1972a, c], is an operational approach to cephalic morphogenesis. Based on a synthesis of experimental,
clinical, and conceptual data, the method strongly supports the thesis that
skeletal organ growth regulation reflects, temporally and ontogenetically,
secondary, compensatory, and mechanically obligatory responses to the
prior and morphogenetically primary demands of those non-skeletal tissues, organs, and functioning spaces that themselves actually carry out
the several (and integrated) functions of the head and neck. Although this
method has been applied principally to studies of rodent and human cephalic growth, it is, in fact, a specific application of a general concept
equally applicable to all vertebrate cephalogenesis, to vertebrate post-
cranial skeletal growth [Moss, 1972b], and to certain aspects of in-
vertebrate skeletal growth [Moss and MEEHAN, 1968]. A specific analysis
of primate craniofacial growth is presented here for the first time utilizing
this method. It is appropriate to review first certain general principles.

Principles of Functional Cranial Analysis

The head (and neck) is a region of the body within which certain
functions are carried out: digestion, respiration, vision, olfaction, hearing,

1   Aided in part by NIDR grant DE-00132.

balance, neural integration, etc. Each separate function is carried out completely by a *functional cranial component* (FCC), i.e., one component for digestion, one for vision, another for olfaction, etc. Each component in turn, consists of two parts. The first is all of the organs, tissues, blood vessels, nerves, and functioning spaces (oral, nasal, pharyngeal) necessary to carry out *completely* a given function and are termed the *functional matrix* (FM) of the specific FCC. The second consists of all of the skeletal tissues (bone, cartilage, tendon, etc.) required to provide biomechanical protection and/or support of its specific FM. This is termed the *skeletal unit* (SkU). In a type of formulation we have:

FCC = FM + SkU.

Several points require clarification. Obviously, any FM is a constellation whose several parts may (or may not) also be operationally involved in other functions. Their roles in several functions are not mutually exclusive. Each function, and hence each FCC, has its unique constellation of parts forming its specific FM. Entirely similar statements are possible for the SkU. The classically denominated bones of formal osteology are not in any sense the equivalents of SkU. Virtually all bones studied reveal that the size and shape of any given skeletal unit is not *necessarily* (in the sense of biological causation) correlated to the form (where form = size and shape) of any other skeletal unit. Further, the form, as well as the spatial position of any given FCC, is relatively independent of these same attributes of any other FCC (see Moss and YOUNG, 1960; Moss, 1968; Moss and SALENTIJN, 1969a, for details and further citations).

All skeletal tissues (and organs) arise, grow, and are maintained in being completely embedded within their specific functional matrices. Following the initial ontogenesis of skeletal tissues, a substantial body of data strongly supports the view that neither the form nor the spatial location of any SkU is an expression of genetic activity operating directly at the level of the scleroblastic cell. Specifically, it is denied that any of the morphological attributes of skeletal tissues beyond specific cytodifferentiation are produced directly by the operation of encoded genetic information within the nuclei of any one or any group of scleroblastic cells (see Moss, 1972b, for a recent review). Rather, the *in vivo* size and shape, growth, and maintenance of skeletal tissues and organs (as SkU) are an indirect and secondary expression of genetic activity of the related functional matrices. These matrices, in turn, may well have significant portions of their genetic expression neurotrophically regulated, in addition to whatever general systemic or specific extrinsic factors act to regulate the several matrices [Moss, 1971a, 1972c].

Traditionally, a triad of processes has been held responsible for the totality of craniofacial growth, the emphasis upon one or another aspect of the triad varying with the student. The three are: (1) sutural tissues, acting as analogues of growth plates, act as primary centers of expansive growth, literally moving bones apart; (2) cephalic cartilages (condylar, basal, nasal) similarly act as sites of primary expansive growth; (3) osseous deposition and resorption act to alter bone form and *pari passu* relocate it in space. The first statement is demonstrably wrong, as data on all available sutural areas show; the second, to date, has been shown to be incorrect for the nasal and condylar cartilages [Moss and BROMBERG, 1968; Moss and RANKOW, 1968]. Growth at both sites is secondary and compensatory. The third is correct, but is deficient in magnitude in some areas and often opposite in direction to the observable diametric growth [Moss *et al.*, 1972].

This situation is resolved by the recognition of two types of functional matrices, each associated with a distinctive type of growth process. We differentiate between *periosteal* and *capsular* matrices. The former is exemplified by a skeletal muscle, typically attaching to the fibrous periosteum; the latter derives its use from the fact that cephalic periosteal FM and their specific SkU arise, grow, and exist within either a neurocranial or an orofacial capsule [see Moss and SALENTIJN, 1969a]. Alterations in the functional demands of periosteal matrices act directly upon their SkU. They do so by the processes of deposition (or formation) and removal (i.e., by all of the histologically observable processes usually associated with the 'growth' of all skeletal tissues), producing changes in SkU form only. This is termed *transformative* growth. The expansion of cephalic capsules (as responses to the volumetric growth of enclosed and protected neural or orbital FM masses, for example, or as responses to the mitotic activity of the epithelially-covered and -lined orofacial capsules) causes all of the embedded SkU to be *passively* moved in space. Such spatial relocations are not accomplished by the histologically observable 'active' growth processes mentioned above, but they do accomplish growth movements of major dimensional magnitude [Moss and SALENTIJN, 1969b]. This is *translative* growth. Both processes can occur either simultaneously or consecutively. The sum of transformation and translation is the totality of growth. Until recently, translative processes were not comprehended; and all studies were related to transformative growth alone. It is clear now that during active transformative growth one of the classic triad of growth processes is involved (deposition and resorption); while passive translative growth is a primary morphogenetic event that can be followed directly by compensatory growth (accretions) at both sutural margins and in cephalic cartilages [see Moss and SALENTIJN, 1969b, for comprehensive citations].

With the statement of these general principles, we may now turn to the analysis of craniofacial growth in primates.

The Applicability of the Method of Functional Cranial Analysis to the Primate Skull

The literature on primate cranial growth, while somewhat voluminous, is at least one order of magnitude less than that dealing with man. Nevertheless, the reader will appreciate that it is beyond the scope of this paper to present a comprehensive bibliography. However, reference to the citations provided will lead quickly to the core of available material.

In recent years, students of primate cranial growth increasingly recognize the value of considering the skull in terms of its functional components [VAN DER KLAAUW, 1948–1952; Moss, 1971b]. It is true that not all workers accept the basic premise of VAN DER KLAAUW and myself that it is both correct and valuable to consider the head in this 'atomistic' fashion. There is another viewpoint, the 'holistic', which insists that the morphological and functional *integration* of the several cranial components must be considered if a correct view is to be achieved [DULLEMEIJER, 1971]. Without entering into the complexity of the discussion at this point [see Moss, 1969], it will suffice to appeal to the law of scientific parsimony here to state that more meaningful experiments may be constructed on an atomistic basis than on a holistic one. Some of the complexities that inevitably arise when the holistic approach is used are exemplified in DU BRUL [1965]. Many other workers have successfully followed the atomistic method: TUCKER [1957] speaks of 'functional units'; VOGEL [1966] analyzes 'functional complexes'; ZINGESER [1966] speaks of 'functional matrix'; and DUTERLOO and ENLOW [1970] now state that 'two basic types of growth movement are involved in skeletal enlargement', the first corresponding to the active, transformative processes and the second, a 'displacement', being the equivalent of the passive translative growth of our concepts. It is reasonable to presume that it is not necessary now to urge consideration of a totally new doctrine or methodology, but rather to provide additional selected examples of the general principles of functional cranial analysis applicable to primate cranial growth. The method chosen here is critical citation of the available literature. The majority of these papers deal with circumscribed topics; while a few provide a general overview [cf. BIEGERT, 1957; HOFER, 1965].

A. Prenatal Development

Available studies of embryonic and early fetal development of the primate head tend to be almost purely descriptive and deal extensively with argumentation concerning the fetalization hypothesis of Bolk [STARCK, 1959, 1961; STARCK and KUMMER, 1962]. Surprisingly, these same authors have been in the forefront of zoologists and comparative anatomists who have stressed the operational and conceptual importance of considering the independence of the several 'functional cranial components' (our term) in the head. The ability to deal with early mammalian cephalogenesis successfully in terms of our concepts has already been demonstrated [SALENTIJN and MOSS, 1971].

B. Neurocranium–Splanchnocranium

There is no necessary (biologically causal) correlation between either the size or spatial position of these two relatively independent skull components (fig. 1). In this figure, with neurocrania reduced to similar (two-dimensional) forms, the splanchnocrania exhibit wide variability. This point is strongly reinforced by FRICK [1960] (fig. 2). Here we observe the variability of prebasal kyphoses in two adult male *Papio cynocephalus*, both collected in the same Ethiopian locality.

C. The Neural Skull

The neurocranium as a whole, or, better, the neural capsule, exists to protect and support the enclosed neural mass (brain, cerebrospinal fluid, nerves, vessels, leptomeninges). The growth of the neural mass is not an event 'permitted' by the prior growth of neural skeletal elements; the situ-

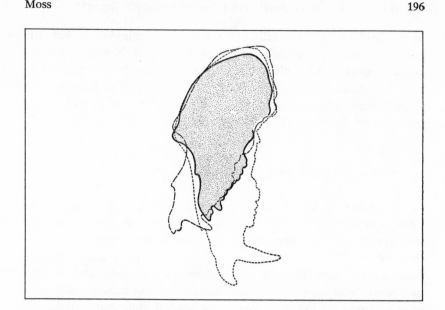

Fig. 1. Three primate skulls, in *norma lateralis*, are reduced to essentially similar endocranial outlines and registered on the cranial base. The figure shows clearly
the lack of correlation between the size and position of the neurocranium and the
size and position of the splanchnocranium. Solid line: *Theropithecus gelada*; interrupted line: *Papio hamadryas*; dashed line: *P. cynocephalus* [adapted from HOFER,
1965].

ation is precisely the opposite. The growth of the neural mass is the primary (and prior) event in expansion of the neural capsule and therefore
accounts for the passive translation of all enclosed SkU and periosteal
FM [cf. STARCK, 1953].

The single calvarial bones (e.g., the parietal) are not unitary functional elements; rather, their ectocranial, endocranial, and diploic (or
pneumatic) regions are individual skeletal units responsive to differing
functional matrices. There can be little doubt that cranial cresting in primates reflects a secondary, compensatory response to altering demands of
specific periosteal matrices [ASHTON and ZUCKERMAN, 1956b; HOLLO-
WAY, 1962]. In a more sophisticated statement, VOGEL [1962] has analyzed the developmental relationship between the temporalis muscle and
its skeletal unit (the coronoid process and sagittal crest).

The general pattern of endocranial growth (at least for the extant
apes) is similar to that observed in man [ASHTON and SPENCE, 1958];
while, as expected, the growth changes of position of the foramen magnum

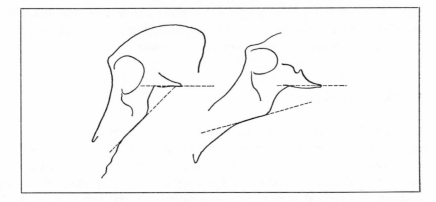

Fig. 2. The variation of prebasilar kyphosis between two male *Papio cynoce-phalus,* derived from the same geographic location, is shown. The lack of any necessary (or causal) relationship between the spatial positioning of neural and facial skulls is evident [adapted from FRICK, 1960].

differ [ASHTON and ZUCKERMAN, 1956a]. In those animals having relatively small body (or splanchnocranial) size and relatively large brain (neural mass) size, the general configuration of the skull will be influenced by the neural mass. However, in other animals the configuration of the endocranium alone (as skeletal units) responds to the functional demands of the neural mass [HOFER, 1954a, b]. There is a relatively large literature on this problem (compare the hyena and man, for example). Indeed, the differences between the neurocranial growth of the thick-skulled pig and of the thin-skulled rat furnished several decades ago one of the now classical grounds for misunderstanding of the general principles of neurocranial growth (the old arguments among BRASH, MASSLER, and SCHOUR are reviewed in MOSS, 1954; see also HOFER, 1954a, b).

The correctness of considering the pachymenix as a skeletal element in the neurocranium [MOSS, 1960] is supported for primates by HOFER [1954a, b]. A proper functional analysis of pneumatic cranial sinuses has not yet been completed in any animal.

WEGNER [1956] at least provides a comprehensive review of the anatomical data on the paranasal sinus. The development of the tympanic bulla [VON SPATZ, 1966] and of the mastoid air cells [SABAN, 1964] again give us data which may serve in the future to illustrate the principles of functional cranial analysis as they apply to the responses of specific skeletal units responding to the needs of the aural functional matrix. The patterns of calvarial (and facial) sutural closure with age in primates, while showing expected specific differences, do not differ from the general principles already enunciated for mammals as a whole [cf. CHOPRA, 1957].

D. Orbital Region

The thesis that the growth of the bony orbit as a skeletal unit responds to the developing orbital mass (as a functional matrix) has been presented [MOSS and YOUNG, 1960]. Recently, EHARA and SEILER [1970] have gone on to show that several additional ectocranial markings of the primate supraorbital region represents morphogenetic responses to additional demands of muscular (periosteal) FM. The increasing importance of binocular vision as one progresses from insectivores to primates should, according to the functional concepts espoused here, have morphogenetic consequences on cranial form and position. Such data are presented by VON SPATZ [1968, 1970]; these effects may be observed either in the neural or in the facial skull (see also DU BRUL, 1965, for a unique argument).

E. Facial Skull

Most studies of facial growth in man employ either (a) anthropometric or craniometric techniques (used by anatomists, physical anthropologists), or (b) cephalometric roentgenography (used by orthodontists; see Moss, 1971a, b, for a critical review). With regard to nonhuman primates, a similar dichotomy of approach and consequent data exists. VOGEL [1966] cites most of the relatively scarce literature to that date.

One group of reports provides essentially biometric data on dimensional or angular growth changes [ASHTON and ZUCKERMAN, 1958]. BIEGERT [1956] deals monographically with the temporomandibular joint. Useful as these are intrinsically, their further significance becomes apparent only when they can be used with a functional viewpoint in further studies. A second group of workers provides us with information directly applicable to a functional analysis of facial growth [ZINGESER, 1966]. ENLOW [1963, 1966], in particular, provides most useful data on the sites and directions of transformative growth processes. Ongoing studies of nonhuman primate cranial growth, using intraosseous metallic implants

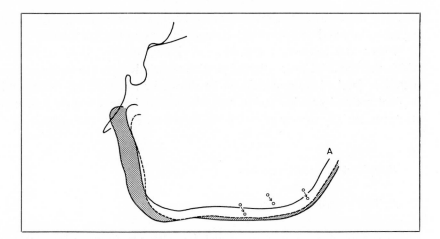

Fig. 3. A demonstration of the relative extent of active (transformative) and passive (translative) growth in a macaque. The original X-rays were supplied by Dr. ROBERT MOYERS, University of Michigan. The technique is that of MOSS and SALENTIJN [1969a, b]. The amount of vertical growth in the period studied due to active (transformative) processes is indicated by the stippled area. The remaining distances, shown by the non-stippled areas, represent passive (translative) growth. Compare these magnitudes with those indicating the displacement of three metallic implant markers. The dominance of passive growth is similar to that seen in man, although the specific amounts, sites, and directions obviously differ.

and cephalometric roentgenographic techniques, are underway at the Center for Human Growth and Development at the University of Michigan and should shortly provide much valuable data.

In certain respects, many of the primary postulates of functional cranial analysis have been well demonstrated in the primate splanchnocranium. VOGEL [1965], dealing with facial sutural morphology, considers that these sutures, like their calvarial counterparts, reflect the local mechanical demands placed upon them. DELLINGER [1967] has shown experimentally that it is possible to alter the dento-alveolar functional cranial component.

The principle thesis that passive translative growth of the mandible is followed by secondary and compensatory changes in the condylar cartilages has been demonstrated repeatedly [JOHO, 1968]. Indeed, it was the work of SARNAT [1971] that clearly demonstrated in primates that following bilateral condylectomy, normal vectors of passive translative growth of other mandibular SkU occurred and, further, that the concur-

rent new demands of the intact temporalis muscles produced corresponding secondary and compensatory changes in the coronoid processes.

Moss and SALENTIJN [1969b, 1971] have published a method by
which the vectorial properties of passive translative growth of the human
mandible (and of the maxilla) can be shown by use of either the metallic
implant method or other stable vital osseous landmarks. Figure 3 demonstrates essentially similar growth processes in the primate mandible.

Supplementary Approaches

A. The Vestibular Method

A school of craniology exists at Lille, whose methods, principles, and
conclusions generally are not widely known. Since much of their work
deals with primates, it is appropriate to mention it here. It is based essentially upon a sound physiological base, i.e., that the horizontal semicircu-
lar canal is always precisely in that orientation *in vivo* [DELATTRE and
FENART, 1958a]. Among their voluminous papers we note here several of
importance for our present concerns [DELATTRE and FENART, 1958b,
1963]. A specific study of anthropoid cranial ontogenesis is of particular
interest [DELATTRE and FENART, 1956a], as is their study of the growth of
the primate splanchnocranium [DELATTRE and FENART, 1956b]. A generalized monographic study is also available [DELATTRE and FENART, 1960].
Additional references to this type of work and its application to primate
facial growth is found typically in the recent work of DARDENNE [1970].

The first value of this approach is that it offers an alternative of
greater biological validity to the usual methods of orientation and registration common to orthodontically-directed studies that use the non-tradi-
tional techniques of cephalometric roentgenography. The second and perhaps more important value is that it permits the development of some
rather new concepts of craniofacial ontogenesis (and phylogenesis). Despite the fact that the Lille authors study the transformative and translative growth of 'bones' (in the classical sense) as a whole, they also are
concerned with and study the growth of several of the FM involved. For
example, as we stressed previously the essential differences between the
ecto- and endocranial surfaces of calvarial bones (as separate SkU), so
they stress the differentiation between ecto- and endofacial osseous surfaces and their differing behavior during growth.

An instructive example of the value of the vestibular method is offered by a comparison of neurocranial growth of nonhuman primates as viewed by DUTERLOO and ENLOW [1970] and by DELATTRE and FENART [1960]. First, let it be noted that both sets of workers agree that, while the same types of general processes are involved in human and primate cranial growth, there are rather great differences in sites, directions, and amounts of growth between these two groups.

DUTERLOO and ENLOW [1970], appropriately concentrating on descriptions of deposition and resorption of osseous surfaces (and giving at least passing acknowledgement to the principle of passive translation, i.e. 'displacement'), observe correctly that the direction of the foramen magnum differs markedly between *Homo* and *Macaca*. They conclude that 'these same remodeling differences, also, are apparently associated with the greater downward angulation of the cranial base flexure in man as compared with the monkey. This, in turn, is related to the more upright posture of the human skeletal frame and the vertical disposition of the facial complex.' One might, therefore, interpret the differences in growth patterns as reflecting relative degrees of basal kyphosis, at least in part. What one gains from the DUTERLOO and ENLOW [1970] study is not any feeling that diametrically opposite growth processes are involved, but rather that it is a question of the relative 'amount' or 'degree' of the same processes.

DELATTRE and FENART [1960], on the other hand, suggest that distinctly different growth processes are involved, as ascertained by their vestibular orientation method. Considering the skull in *norma lateralis* and viewed from the left side, they conclude that during human cranial ontogenesis the posterior portion of the neural skull undergoes a clockwise rotation in man, but a counterclockwise rotation in nonhuman primates. The reader is referred to their work for observations on other aspects of primate cranial growth. Much of it presents new concepts; and all of the observations, we would add, are compatible with the concepts of functional cranial analysis.

B. Cranial Base Flexion

No study of primate cranial growth can ignore the cranial base, which is worthy of monographic presentation elsewhere. It will suffice here to indicate some of the scientific problems this area presents. To some extent the validity of *all* methods of cranial growth study depends upon the particular method of registration and orientation of their graphic data. In addition to the several modes commonly used by orthodontically-trained workers and to the vestibular method discussed above, several German workers have devoted much thought to this problem. Basically recognizing that neurocranial form and position reflect the position and form of the protected and supported portions of the neural mass, they have made a compelling case that holds that a clival registration (and horizontal orientation) is by far the best to use in study of comparative cranial form as well in ontogenetic studies (see HOFER, 1965, for a review and pertinent citations). This subject is certainly worthy of further consideration.

Conclusion

A Functional Cranial Analysis

In this concluding section it is well to summarize our material and to offer a conceptual synthesis. For obvious reasons a great deal of the experimental work underlying the method of functional cranial analysis was performed upon non-primate laboratory animals, the most notable exception being the use of small primates to study the role of the mandibular condylar cartilages and to carry out ongoing *in vivo* implant studies. Despite the lack of specific experimental data, it is possible to reach some valid conclusions.

There can be little doubt that the general characteristics of the growth processes of all other mammalian skulls are equally valid for all primates. Primates as a group, as well as species and individuals, may and probably do exhibit differences in rate, direction, amount, and sites of growth. At this time it is true that we do not yet possess sufficient data to delineate such differences accurately for any but a selected few nonhuman primates, those most capable of scientific colonization. With these points in mind we may now summarize.

The neural skull. The primary morphogenetic agency of the neurocranium is the volumetric increase of the neural mass, enclosed within the neurocranial capsule. Further, the very induction of the basicranial cartilages is an interaction between the several portions of the brain stem and responding ectomesenchymal cells. Expansion of the neural mass causes the neurocranial capsule to expand. There are data suggesting that the initial sites of intramembranous calvarial ossification are related to biomechanical (and inductive) forces, produced within this capsule by the neural mass and distributed by the organized tracts of the fibrous dura mater, which together with the base and calvaria form a biomechanical unity [Moss, 1960]. This capsular functional matrix may interact with the orbital capsular matrix (as well as the dural) to produce eventual neurocranial form.

As the calvarial bones are passively translated outwards, transformative growth processes occur simultaneously. These produce secondary, compensatory growth at sutural margins, as well as the differential responses of ecto- and endocranial SkU to different FM (muscles, gyri, blood vessels, etc.). Differential growth of portions of the neural mass (both dimensional and angular) occurs with consequent secondary transformative and translative changes of SkU. Alterations of basal kyphosis are not morphogenetic events originating in the several basicranial bones and cartilages. Rather, they are secondary responses to primary changes in either (or both) neural or nasopharyngeal matrices. Similarly, there are no experimental data to support the view that basal synchondroses are sites of intrinsically regulated and morphogenetically primary expansive growth. All available data fit equally well with the concepts, based firmly on experimental work on the nasal and mandibular condylar cartilage, that these basal synchondroses also are sites of secondary, compensatory growth. The reader's attention is directed to the critical distinction between the limited ability of cartilaginous tissues *per se* to elongate linearly for brief periods of time under experimental conditions and the role of cartilaginous tissues as portions of skeletal organs where continued linear growth can be shown to depend upon extrinsic (matrix) effects [see Moss, 1972b].

Calvarial sutural areas are also sites of secondary and compensatory growth. The location of any sutural area and, therefore, the extent (area) of any calvarial bone is not intrinsically determined [Moss, 1954]. Further, all aspects of individual sutural morphology reflect only extrinsic (functional matrix) demands placed upon it [Moss, 1960]. The marked differentiation of the ectocranial contours of many primate skulls accurately reflects differing periosteal matrix demands, many of which show individual variation as well as sexual dimorphism.

The facial skull. Differential spatial positioning of the neural and facial skulls reflects primarily differing positioning of neural and facial functional matrices (see Moss and SALENTIJN, 1971, for a human clinical example). The primary morphogenetic agency in facial growth is the volumetric expansion of the oro-naso-pharyngeal functioning spaces. This is not to be conceived of as any form of simplistic pneumatic force, but rather as a neurotrophically-regulated epithelial growth process. There is little reason to believe, additionally, that the neurally-regulated airway maintenance mechanism postulated by BOSMA [1963] does not function equally in nonhuman primates (as in all mammals). Mid-face growth is not brought about by any primary expansive force intrinsically resident in the nasal septal cartilage; nor is the lowering and anterior positioning of the mandible the result of a similar primary force generated within the mandibular cartilages (which, in fact, are articular structures and not in any way homologous with the growth plates of long bones). Expansion of the several orofacial capsular matrices causes the orofacial capsule to expand in a manner entirely homologous to the neurocranial capsule. The midfacial membranous bones respond homologously to the calvarial bones, undergoing passive translative, as well as active transformative growth. The lower face grows similarly. Passive translation of the mandible is followed by secondary, compensatory growth at the condylar cartilages, effectively keeping the temporomandibular joint in function. Since periosteal matrices are also within the expanding orofacial capsule, it is not surprising that these, in turn, produce the transformative growth changes (of the several mandibular ramal SkU, for example) that are so well known.

Finally, although this point has not been discussed in this paper and few data are yet at hand, it is reasonable for this laboratory to presume that nonhuman primate orofacial growth is indeed capable potentially of expression as a function of some logarithmic spirals describing the course of the trigeminal nerve branches and that orofacial growth in primates is equally gnomic in character. The illustrations of DELLATRE and FENART [1960] provide at least tentative evidence that our recently presented hypothesis concerning this matter in human growth is equally true for nonhuman primates [Moss and SALENTIJN, 1971; SALENTIJN and Moss, 1971].

Functional cranial analysis, then, provides a conceptual framework within which significant analyses of primate craniofacial growth can be made. The essence of the method is not so much that it provides new data, although it does, but rather that it provides a new way of looking at all data. The basic fact of the diversity of craniofacial growth patterns in nonhuman primates makes these animals useful to test the validity of this, as well as all other concepts of cranial growth. The ability to demonstrate the validity of any given concept against this background of diversity should make our studies of the growth of the relatively less variable human head more nearly approach the truth.

References

ASHTON, E. H. and SPENCE, T. F.: Age changes in the cranial capacity and foramen magnum of hominoids. Proc. zool. Soc., Lond. *130:* 169–181 (1958).

ASHTON, E. H. and ZUCKERMAN, S.: Age changes in the position of the foramen magnum in hominoids. Proc. zool. Soc., Lond. *126:* 315–325 (1956a).

ASHTON, E. H. and ZUCKERMAN, S.: Cranial crests in the anthropoidea. Proc. zool. Soc., Lond. *126:* 581–634 (1956b).

ASHTON, E. H. and ZUCKERMAN, S.: The infraorbital foramen in the hominoidea. Proc. zool. Soc., Lond. *131:* 471–485 (1958).

BIEGERT, J.: Das Kiefergelenk der Primaten. Morph. Jb. *97:* 249–404 (1956).

BIEGERT, J.: Der Formwandel des Primatenschädels und seine Beziehungen zur onto-genetischen Entwicklung und den phylogenetischen Spezialisationen der Kopf-organe. Morph. Jb. *98:* 77–199 (1957).

BOSMA, J.: Oral and pharyngeal development and function. J. dent. Res. *42:* 375–380 (1963).

CHOPRA, S. R. K.: The cranial suture closure in monkeys. Proc. zool. Soc., Lond. *128:* 67–112 (1957).

DARDENNE, J.: Etude comparative des principaux paramètres sagittaux de la face et du crâne, chez l'homme et les chimpanzés par la méthode vestibulaire d'o-rientation; Thèse Doct. Chriurg. Dent., Univ. de Lille (1970).

DELATTRE, A. et FENART, R.: Analyse morphologique du splanchnocrâne chez les primates et ses rapports avec le prognathisme. Mammalia *20:* 169–325 (1956a).

DELATTRE, A. et FENART, R.: Etude de l'ontogénèse du crâne des anthropoids du Congo Belge. Ann. Mus. roy. Congo Belge *47:* 11–121 (1956b).

DELATTRE, A. et FENART, R.: La méthode vestibulaire. Z. Morph. Anthrop. *49:* 90–114 (1958a).

DELATTRE, A. et FENART, R.: Essai de systématisation du pariétal et son utilization au cours de l'étude de sa croissance. Bull. Soc. Anthrop. *10:* 245–295 (1958b).

DELATTRE, A. et FENART, A.: L'hominisation du crâne (Centre nat. Rech. Sci., Paris 1960).

DELATTRE, A. et FENART, A.: Etude des projections horizontales et vertico-frontales du crâne au cours de l'hominisation. Anthropologie *67:* 85–114, 301–346, 525–563; *68:* 95–132 (1963).

DELLINGER, E. L.: A histologic and cephalometric investigation of premolar intru-sion in the *Macaca speciosa* monkey. Amer. J. Orthod. *53:* 325–355 (1967).

DU BRUL, E. L.: The skull of the lion marmoset *Leontideus rosalia* Linneaus: a study in biomechanical adaptation. Amer. J. phys. Anthrop. *23:* 261–276 (1965).

DULLEMEIJER, P.: Comparative ontogeny and cranio-facial growth; in MOYERS and KROGMAN Cranio-facial growth in man, pp. 45–75 (Pergamon Press, Oxford 1971).

DUTERLOO, H. S. and ENLOW, D. H.: A comparative study of cranial growth in Homo and Macaca. Amer. J. Anat. 127: 357–368 (1970).

EHARA, A. und SEILER, R.: Die Strukturen der Überaugenregion bei den Primaten. Deutung und Definitionen. Z. Morph. Anthrop. 62: 1–29 (1970).

ENLOW, D. H.: Principles of bone remodeling (Thomas, Springfield 1963).

ENLOW, D. H.: A comparative study of facial growth in Homo and Macaca. Amer. J. phys. Anthrop. 24: 293–307 (1966).

FRICK, H.: Über die Variabilität der präbasialen Kyphose bei Pavianschädeln. Z. Anat. EntwGesch. 121: 446–454 (1960).

HOFER, H.: Beobachtungen am Hirnrelief der Aussenfläche des Schädels, am Endokranium und der Hirnform des südamerikanischen Nachtaffen (Aotes, Ceboidea). Ber. oberhess. Ges. Nat.-Heilk. 27: 90–110 (1954a).

HOFER, H.: Die cranio-cerebrale Topographie bei den Affen und ihre Bedeutung für die menschliche Schädelform. Homo 5: 4–45 (1954b).

HOFER, H.: Die morphologische Analyse des Schädels des Menschen; in HEBERER Menschliche Abstammungslehre, pp. 145–226 (Fischer, Stuttgart 1965).

HOLLOWAY, R. L., jr.: A note on sagittal cresting. Amer. J. phys. Anthrop. 20: 527–530 (1962).

JOHO, J. P.: Changes in form and size of the mandible in orthopaedically treated Macacus irus (an experimental study). Trans. europ. orthodont. Soc. 44: 161–173 (1968).

KLAAUW, C. J. VAN DER: Size and position of the functional components of the skull. Arch. neerland. Zool. 8: 1–559 (1948–52).

MOSS, M. L.: The growth of the calvaria in the rat: the determination of osseous morphology. Amer. J. Anat. 94: 333–362 (1954).

MOSS, M. L.: Inhibition and stimulation of sutural fusion in the rat calvaria. Anat. Rec. 136: 457–468 (1960).

MOSS, M. L.: The primacy of functional matrices in orofacial growth. Dent. Pract. 19: 65–73 (1968).

MOSS, M. L.: A theoretical analysis of the functional matrix. Acta biotheoret. 18: 195–202 (1969).

MOSS, M. L.: Neurotrophic processes in orofacial growth. J. dent. Res. 50: 1492–1949 (1971a).

MOSS, M. L.: Ontogenetic aspects of cranio-facial growth; in MOYERS and KROGMAN Cranio-facial growth in man, pp. 109–124 (Pergamon Press, Oxford 1971b).

MOSS, M. L.: Twenty years of functional cranial analysis. Amer. J. Orthod. (in press, 1972a).

MOSS, M. L.: The regulation of skeletal growth; in Goss The regulation of organ growth (Academic Press, in press, 1972b).

MOSS, M. L.: An introduction to the neurobiology of oro-facial growth. Acta biotheoret. (in press, 1972c).

MOSS, M. L. and BROMBERG, B.: The passive role of nasal septal cartilage in mid-facial growth. Plast. reconstr. Surg. 41: 536–542 (1968).

Moss, M. L. and Meehan, M. A.: The growth of the echinoid test. Acta anat. *69:* 409–444 (1968).

Moss, M. L.; Meehan, M. and Salentijn, L.: Transformative and translative growth processes in neurocranial development in the rat. Acta anat. *81:* 1–22 (1972).

Moss, M. L. and Rankow, R.: The role of the functional matrix in mandibular growth. Angle Orthod. *38:* 95–103 (1968).

Moss, M. L. and Salentijn, L.: The primary role of functional matrices in facial growth. Amer. J. Orthod. *55:* 566–577 (1969a).

Moss, M. L. and Salentijn, L.: The capsular matrix. Amer. J. Orthod. *56:* 474–490 (1969b).

Moss, M. L. and Salentijn, L.: Differences between the functional matrices in anterior open and deep overbite. Amer. J. Orthod. *60:* 264–280 (1971).

Moss, M. L. and Young, R. W.: A functional approach to craniology. Amer. J. phys. Anthrop. *18:* 281–292 (1960).

Saban, R.: Sur la pneumatisation de l'os temporal des primates adultes et son développement ontogénique chez le genre *Alouatta* (Platyrrhinien). Morph. Jb. *106:* 569–593 (1964).

Salentijn, L. and Moss, M. L.: Morphological attributes of the logarithmic growth of the human face: gnomic growth. Acta anat. *78:* 185–199 (1971).

Sarnat, B. G.: Surgical experimentation and gross potential growth of the face and jaws. J. dent. Res. *50:* suppl., pp. 1462–1476 (1971).

Spatz, W. B. von: Zur Ontogenese der Bulla Tympanica von *Tupaia glis* (Prosimiae tupaiiformes). Folia primat. *4:* 26–50 (1966).

Spatz, W. B. von: Die Bedeutung der Augen für die sagittale Gestaltung des Schädels von *Tarsius* (Prosimiae tarsiiformes). Folia primat. *9:* 22–40 (1968).

Spatz, W. B. von: Binokuläres Sehen und Kopfgestaltung. Ein Beitrag zum Problem des Gestaltwandels des Schädels der Primaten, insbesonders der Lorisidae. Acta anat. *75:* 489–520 (1970).

Starck, D.: Morphologische Untersuchungen am Kopf der Säugetiere, besonders der Prosimier. Ein Beitrag zum Problem des Formwandels des Säugerschädels. Z. wissensch. Zool. *157:* 169–219 (1953).

Starck, D.: Das Cranium eines Schimpansenfetus (*Pan Troglodytes* [Blumenbach 1799]) von 71 mm Schstlg., nebst Bemerkungen über die Körperform von Schimpansenfeten (Beitrag zur Kenntnis des Primatencraniums II). Morph. Jb. *100:* 559–647 (1959).

Starck, D.: Ontogenetic development of the skull of primates. Int. Colloq. on Evolution of Lower and Non-Specialized Mammals. Kon. VI. Acad. Wet. Lett. Sch. Kunst, België; part I, pp. 205–214 (Brussels, 1961).

Starck, D. und Kummer, B.: Zur Ontogenese des Schimpansenschädels (mit Bemerkungen zur Fetalisationshypothese). Anthrop. Anz. *25:* 204–215 (1962).

Tucker, R.: Shape and function in present day craniology. Proc. Zool. Soc., Calcutta, Mookerju Memor. Vol., pp. 207–221 (1957).

Vogel, C.: Untersuchungen an Colobus-Schädeln aus Liberia unter besonderer Brücksichtigung der Crista sagittalis. Z. Morph. Anthrop. *52:* 306–332 (1962).

Vogel, C.: Über den Nahtverlauf im Gesichtsskelett Catarrhiner Primaten einschliesslich des Menschen. Anthrop. Anz. *29:* 301–312 (1965).

VOGEL, C.: Morphologische Studien am Gesichtsschädel Catarrhiner Primaten. Bibl. primat. No. 4, pp. 1–226 (Karger, Basel 1966).

WEGNER, R. N.: Studien über Nebenhöhlen des Schädels. Verschmelzungen der Nebenhöhlen der Nase und Abänderungen ihrer Öffnungen bei Primaten. Wissenschaft. Z. Ernst-Moritz-Arndt-Univ., Greifswald. Mathem. Naturwissenschaft, Reihe Greifswald. 5: 1–111 (1956).

ZINGESER, M. R.: Occlusofacial relationships in the mature howler monkey. Amer. J. phys. Anthrop. 24: 171–180 (1966).

Author's address: Dr. MELVIN L. MOSS, Department of Anatomy, Columbia University, 630 West 168th Street, New York, NY 10032 (USA)

Symp. IVth Int. Congr. Primat., vol. 3: Craniofacial Biology of Primates,
pp. 209–226 (Karger, Basel 1973)

The Monkey Facial Skeleton after Postnatal
Surgical Experimentation[1]

B. G. SARNAT

Department of Plastic Surgery, Division of Surgery, Cedars-Sinai Medical
Center and Research Institute; and Division of Oral Biology, School of
Dentistry, University of California, Los Angeles

Introduction

A. Bone Biology and Facial Growth

Some principles of the biology of bone as they apply to the primate
face are central for this selective review and summary and are, therefore,
briefly discussed in this section. The three modes of postnatal growth and
changes of bones are (1) cartilaginous, (2) sutural (in the skull), and (3)
appositional and resorptive (remodeling) [SARNAT, 1971a]. In the growing
monkey, skeletal mass increases because cartilaginous and sutural
growth are active and apposition is greater than resorption. In the adult,
skeletal mass is constant because apposition and resorption, although active, are in equilibrium; while cartilaginous and sutural growth have
ceased. In old age, skeletal mass decreases because resorption is more active than apposition. The skull, a complex of bones, has proved to be a
productive source of study, since in no other part of the body are all of
these three types of postnatal bone growth represented. As a result of the
harmonious activities of these processes at the respective growth sites in

1   This research was supported in part by research grants HD 00179, National Institute of Child Health and Human Development, and RR 05468, US Public Health
Service.

the skull, the normal facial skeleton increases in size in all three dimensions at different times and at different rates [WEINMANN and SICHER, 1955; ENLOW, 1968; SARNAT, 1971b].

Sites of cartilaginous growth in the trunk and limbs are the epiphyses of the long or tubular bones, the costochondral junction of ribs, and the clavicle. The skull growth at the sphenoethmoidal and sphenooccipital synchrondoses, the septoethmoidal and septopresphenoid joints, and the mandibular condyle is in each case cartilaginous, although the histologic arrangement of the adjacent tissues is not the same [ROY and SARNAT, 1956]. These regions and the nasal septal cartilage play a role in the downward and forward growth and movement of the face [WEINMANN and SICHER, 1955; SCOTT, 1967; ENLOW, 1968; SARNAT, 1970].

Growth of cranialfacial bones is also active at sutures, which are found only in the skull. The adult form of the bones represents in part the past pattern of activity, often variable, at the sutures. When sutural growth was measured either on the skull or on serial radiographs, its true direction was difficult to observe, not only because the various sutures grew at different rates at different times, but also because their spatial position in the skull varied. It is likely that with growth the sutural planes change in relation to each other. Growth of the face thus does not follow straight lines. Rather, with the rotation of the sutural planes, the bones of the face follow various curves. Measurements of this growth on the radiographs, however, could show only linear change.

A basic physiologic concept is that throughout life bone is in a continuous state of apposition and resorption. Consequently, through remodeling, skeletal form is always subject to change. The basic and dual response of continuous bony apposition, as well as modeling resorption, occurs along the highly sensitive periosteal and endosteal surfaces of bones and contributes to dimensional changes in all directions. The physiologic stability of the bony components is the result of many interrelated factors, normal functional use being a prominent one.

In addition, the contents of certain cavities of the skull influence the different types of growth of adjoining bones and sutures; examples are the relationship of the brain to the neurocranium, the orbital contents to the orbit, and the tongue to the oral cavity [SARNAT, 1971b]. In addition, growth of the cartilaginous nasal septum may influence sutural growth and contribute to growth of the nose and the downward and forward growth of the face and palate with accompanying increase in the size of the nasal cavity [SARNAT, 1970].

B. The Experiments

Surgical experiments were designed to permit assessment of both normal and abnormal facial growth changes of the nasomaxillary complex and the mandible, including the study of normal facial sutural growth by means of the implantation of radiopaque dental silver amalgam followed by gross measurements made directly on the skull and indirectly on serial cephalometric radiographs [GANS and SARNAT, 1951]. The effect of extirpative surgical procedures on facial growth was also studied after resection of the median and transverse palatine sutures in growing monkeys [SARNAT, 1958] and after mandibular condylectomy in growing [SARNAT and ENGEL, 1951; SARNAT, 1957] and adult [SARNAT and MUCHNIC, 1971a] monkeys.

Upper Face

A. Use of Implant Markers and Cephalometric Radiography to Assess Normal Growth at Several Facial Sutures

The purposes of this study in *Macaca mulatta* were (1) to compare the relative amounts of normal growth at the frontomaxillary, frontozygomatic, zygomaticotemporal, zygomaticomaxillary, and premaxillomaxillary sutures by means of metallic implants (fig. 1A); (2) to compare the radiographic with the direct method of measuring growth by means of metallic implants; and (3) to study the direction of growth of certain components of the facial skeleton by superposing tracings of serial radiographs [GANS and SARNAT, 1951].

The monkeys were separated into three groups. Those in the youngest group, estimated to be about 8 months old, had complete deciduous dentitions. The intermediate group had four permanent first molars; animals were estimated to be about 18 months old. Members of the oldest group, because of the presence of permanent central and lateral incisors as well as permanent first molars, were believed to be about 24 months old. The experimental period ranged from 7 to 10 months.

1. Method of Implantation and Measurement

The skin overlying the area selected for implantation was shaved and then cleansed [SARNAT, 1968]. An aseptic technique was observed throughout the surgical procedure. Incisions were planned to avoid interference with motor nerve supply lest the experiment be modified by disturbed muscle function. The skin and subcuta-

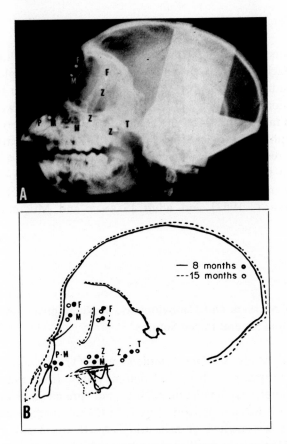

Fig. 1. A Lateral cephalometric radiograph of *Macaca mulatta* skull with radiopaque amalgam implants on each side of frontomaxillary (F-M), frontozygomatic (F-Z), premaxillomaxillary (P-M), zygomaticomaxillary (Z-M), and zygomaticotemporal (Z-T) facial sutures. *B* Superposed tracings of serial cephalometric radiographs. ● = Position of implant at beginning of study; ○ = position of implant
at completion of study. Note downward and forward movement of all implants, except for posterior movement on the temporal side of Z-T [from GANS and SARNAT,
1951].

neous tissues (or mucous membrane) were incised; bleeding was controlled; and the
wound margins were retracted. Underlying bone covered by muscle was freed by
blunt dissection. After exposure, the periosteum was incised, elevated from the
bone, and retracted.

   With an inverted-cone dental bur, undercut cavities were prepared in the corti-
cal plate of bone on both sides of the five sutures to be studied. Dental silver amalgam was packed into the cavities, and an indentation was made with the point of a

caliper in the center of each implant before it hardened. Amalgam was used because of its radiopacity, pliability, and tolerance by tissues. The distance between implants was measured to the nearest 0.1 mm. The soft tissues were then replaced and sutured in layers.

2. Gross Measurement Studies

When the experiment was terminated, gross direct measurements were again made on the dissected skulls to obtain information about total growth at the sutures. Separation of the pair of implants at the zygomaticotemporal suture exceeded that of all other regions studied, followed by that of the implants in the region of the zygomaticomaxillary suture. Separation of paired implants at the frontozygomatic, frontomaxillary, and premaxillomaxillary sutures was considerably less. Variations in the increased distance between paired implants at the same suture in different animals were noted. In several animals, the separation of implants at the frontomaxillary suture exceeded that of the premaxillomaxillary suture; in others the reverse was true. In the groups of older animals the increase in separation was less than in the youngest age group.

Examination of the prepared skulls revealed the prominence of appositional growth along the lateral borders and resorption along the medial borders of the zygomatic arch and zygomatic process of the frontal bone. The amalgam implants placed initially on the lateral surface could no longer be seen from the lateral view; they were now visible only on the medial surface. Because the two methods complement each other, the gross direct measurement method was frequently combined with that of serial cephalometric radiography to study sutural growth of bones.

3. Serial Cephalometric Radiography Combined with Radiopaque Implants

GANS and SARNAT [1951] adapted the Broadbent-Bolton cephalometer for serial cephalometric radiography in combination with metallic implants on either side of several facial sutures in monkeys. This method permitted a longitudinal study and permanent records. The implant images served as references from which information could be obtained as to sites, amounts, rates, and relative direction of growth. In addition, the changes that occurred over time were determined from the serial radiographs without killing or further surgery on the animal. The increase in distance between paired implants was thus measured readily and repeatedly. Cephalometric radiographs (as in fig. 1A) were taken on the Broadbent-Bolton cephalometer at monthly intervals. Tracings of the original and final lateral radiographs were superposed on the outlines of sella turcica and the most superior portion of the anterior cranial fossa (fig. 1B).

Comparison of the two tracings showed a downward and forward movement of all implants except those on the temporal side of the zygomaticotemporal suture, which moved downward and posteriorly. The occlusal plane and floor of the nose descended in parallel planes. In the youngest group the bony profile of the face was shifted anteriorly, maintaining the original outline; while in the intermediate and oldest groups there was a trend toward snouting. Measurements taken between implant images on the radiographs followed closely the gross measurements taken on the specimens.

4. Comment

The zygomaticotemporal and zygomaticomaxillary sutures were related to anteroinferior growth; while the frontomaxillary and frontozygomatic sutures contributed most to the downward growth of the face. Growth at these sutures did not account for the entire vertical growth of the face, however. The floor of the nose descended to a lower level than could be accounted for by growth at the frontomaxillary and frontozygomatic sutures; the same condition was true of the occlusal plane of the teeth. Thus, some other factor, possibly the cartilaginous nasal septum, may have played a role. Although no two animals exhibited identical quantitative growth, the general pattern was similar.

B. Extirpation of the Median and Transverse Palatine Sutures in Growing Monkeys

The purpose of this experiment was to determine grossly the effects of complete unilateral removal of the hard palate, including the median and transverse palatine sutures, upon palatal and facial growth in the otherwise normal monkey [SARNAT, 1958].

Because the growth activity of the face is greatest during early life, the youngest *M. mulatta* obtainable were used. Their dental age at the beginning of the experiment was estimated to be about 8 months; chronologic age was not known. The animals weighed from 4.5 to 6 lb. A total of 12 monkeys was included in this report, 7 unoperated controls and 5 experimental animals. In these 5, the mucoperiosteum was first removed from the left half of the hard palate; and then the exposed left bony palate was resected (with a tapered-fissure dental bur), including the median and left transverse palatine sutures, the major palatine foramen, and the nasal mucoperiosteum. Care was taken not to disturb either the alveolar process or the teeth. Thus, communication was established between the oral and nasal cavities by the surgical cleft. Unilateral surgical procedures were performed to permit comparison of the operated and the unoperated sides in the same animal. The postoperative survival period ranged from 1 to 34 months.

Examination of the palates postoperatively revealed that the surgically-exposed bone was now covered by soft tissue with a smooth epithelial surface devoid of rugae. The surgically-produced clefts of the hard palates, with communication between the oral and nasal cavities, persisted in varying degrees. At *post mortem*, examination of the soft tissue revealed the absence of rugae on the operated side (fig. 2A). In addition, the rugal pattern on the unoperated side was not regular. The form varied, and some of the rugae extended beyond the midline, in contrast to the regular and bilaterally symmetrical rugal pattern in the unoperated animals. The size of the clefts ranged from a narrow slit with overlapping of epithelial-covered tissue (fig. 2A) to an extensive cleft (fig. 2C) including the boundaries of the surgical procedure. In dissection of the soft tissue from the hard palate, the normal mucoperiosteum, which contained the rugae, was readily elevated from the bone. The scar tissue, which had no rugae, was thinner and was separated from the bone with more difficulty. Nothing unusual was noted for the soft palate. In every animal, the extensiveness of the bony palatal defect was masked by the overlying soft tissues (fig. 2); where the palatal defect had been bridged by bone (fig. 2E), an eccentrically – placed suture was found, not in the midline, but rather on the operated side. No definite correlation could be made between the size of the soft tissue or size of the bony cleft at *post mortem* and the length of postoperative survival.

The operated and unoperated sides of the skulls were compared with each other; in addition, skulls of monkeys that were operated upon were compared with the unoperated control monkeys. No significant gross difference was noted in growth and development of the hard palate, maxillary arch, mandibular arch, maxillomandibular relationship (arch size and form, occlusion, and tooth relationships), or total face. Within the limits of this experiment, it was concluded that extirpation of the median and transverse palatine sutures did not produce in either the palate or the face a grossly apparent growth arrest. Thus, it might be assumed either that these sutures do not make an important primary contribution to maxillary growth or that other growth sites adjusted to the altered conditions. Since the jaws were in occlusion at the beginning of the experiment, the mandible may have guided maxillary growth.

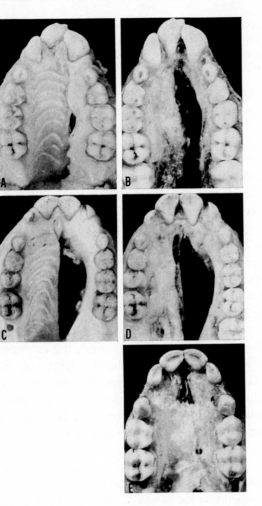

Fig. 2. Post mortem photographs of oral palatal region of monkeys whose left hard palatal regions, including the midpalatine and transpalatine sutures, were completely resected at age of about 8 months. There is no gross asymmetry of the maxillary arch. *A, B* Monkey No. 14, 28 months postoperative survival; upper left permanent central incisor malposed because of absence of adjacent permanent lateral incisor. Note in A the nearly complete soft-tissue closure and lack of rugae on the left side and in B the extensive bony palatal defect and uncovered posterior molars after removal of soft tissue. *C, D* Monkey No. 15, 34 months postoperative survival. Note extensive soft tissue defect and even greater bony tissue defect. *E* Monkey No. 12, 4 months postoperative survival. Note nearly complete bony healing of palatal defect and eccentric suture line toward the operated side [from SARNAT, 1958].

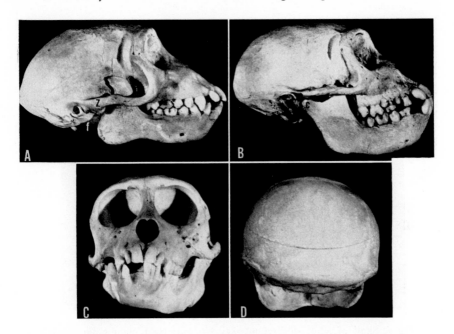

Fig. 3. Skulls of growing rhesus macaques; in A, C, D right mandibular condyle was resected at age of about 8 months; c = condyle; f = false articulation; z = zygomatic arch. *A, C* Lateral and anterior views of animal No. 2–13, 29 months postoperative survival. Facial height is less in A than in B (unoperated animal No. 2–2); no true condyle, fossa, or articular eminence is visible. Note in A considerably wider and heavier coronoid process directed more posteriorly and above zygomatic arch than in B; shorter, wide, anteriorly-positioned ramus; relation of posterior border of ramus to normal locus of true fossa; mandibular angle of less than 90°; accentuation of antegonial notch; and lesser height of mandibular body. In C, note smaller total facial height on operated right side as compared with unoperated left side; relatively lower zygomatic arch; and lesser height of ramus. *D* Posterior view of animal No. 1–6, 14 months postoperative survival. Note higher level of mandible on operated right side (cf. fig. 4) [from SARNAT, 1957].

Changes after Mandibular Condylectomy in Growing and Adult Monkeys

A. Methods and Results

The mandibular condyle was resected on one side in one group of *growing* monkeys *(M. mulatta)* (fig. 3) [SARNAT and ENGEL, 1951; SARNAT, 1957] and another group of *adult* monkeys *(Saimiri sciureus)* (fig. 4)

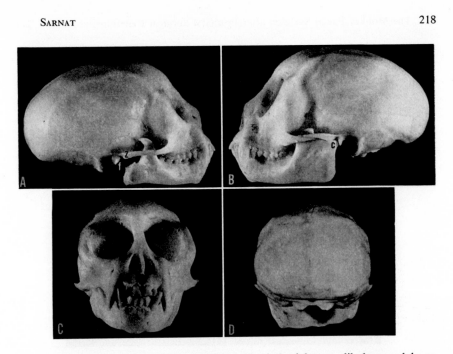

Fig. 4. Skull of adult squirrel monkey No. 3–4; right mandibular condyle resected in adulthood 2 years before death; c = condyle; f = false articulation; z = zygomatic arch. *A, B* lateral views of operated right side and unoperated left side. In A note lesser height; absence of condyle; short posterior ramus and lack of articulation with neurocranium; relatively more prominent coronoid process extending above zygomatic arch; somewhat more prominent antegonial notch. *C, D* Anterior and posterior views, respectively. Note higher level of mandible on operated right side [from SARNAT and MUCHNIC, 1971a].

[SARNAT and MUCHNIC, 1971a]. The gross facial skeletal changes were observed for both groups for as long as two to three years postoperatively. The findings for the operated side, as compared with the unoperated side, were a lesser total facial height and a shorter ramus, with a deficiency greater in the posterosuperior region than the amount of tissue resected (fig. 3 and 4). There was also less maxillary and mandibular alveolar bone, particularly in the molar region, with a lesser height of the zygomatic arch in relation to the lower border of the mandible. The tip of the coronoid process extended well above, rather than to the level of, the zygomatic arch; and the zygomatic arch and external auditory canal were just above the level of the occlusal plane, rather than well above it.

Radiographs were taken of the disarticulated, dissected hemisectioned mandibles of both growing and adult monkeys.[2] The side of the mandible from which the condyle had been resected had no sigmoid notch, a much shorter posterior ramus, and a coronoid process larger relative to the ramus than the unoperated side. The distance from the occlusal level of the teeth to the lower border of the mandible also was less on the operated side and was most apparent in the posterior part of the mandible. The bone of the alveolar crests was flat, and the N-shaped bony trabecular pattern of the ramus, as compared with that of the unoperated side, was disoriented. On the unoperated side, the 'N' of the trabecular pattern consisted of condylar to angular, condylar to retromolar, and coronoid to retromolar area arms. On the operated side, the false condylar to angular arm was shorter; whereas the false condylar to retromolar area arm was longer and not as distinct. There was increased radiodensity of the region of the false condyle and the unoperated condyle. The differences are probably a result of functional changes of the temporomandibular joint with consequent alterations in the direction and amount of muscular pull and the position and motion of the mandible.

In an adult squirrel monkey a tracing of a *post mortem* radiograph of the operated right side (fig. 5B) was superposed in various ways on one of the unoperated left side (fig. 5A) of the mandible to bring out particular findings. When the tracings were superposed on the teeth, an extensive deficiency of bone in the operated side was noted in the superoposterior ramus, superoanterior coronoid, and lower border of the mandible regions (fig. 5C). When superposed on the coronoid processes, surprisingly they coincided (fig. 5D); the anterior part of the operated mandible was rotated downward and the posterior part upward, and the lower borders of the mandibles intersected in the premolar regions. When the tracings were superposed along the lower borders of the mandibles, the deficiency of the ramus of the operated side appeared to be the most extensive (fig. 5E). This superposition also illustrated the amount of bone ultimately lost in relation to the relatively small amount removed two years previously. The anterior border of the coronoid process and the occlusal level of the teeth were lower on the operated side. The areas for superpositioning were selected arbitrarily, and probably no one of them completely reflects the actual relationship.

Tracings were also made of the upper face from *post mortem* photographs of the unoperated left side (fig. 6A) and of the operated right side (fig. 6B) of an adult squirrel monkey. In figure 6C the tracings were

2   These radiographs are figured in SARNAT and MUCHNIC, 1971a, b. [Ed.]

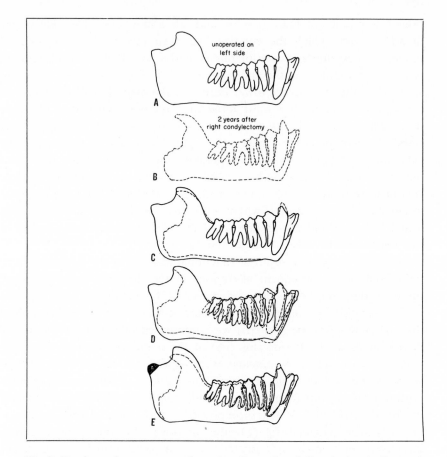

Fig. 5. Tracings of *post mortem* lateral radiographs of disarticulated, dissected, hemisected mandible of adult squirrel monkey No. 3–4; right mandibular condyle resected in adulthood 2 years before death. *A* Unoperated left side (reversed to facilitate comparison). *B* operated right side. *C, D, E* broken line of B superposed on solid line of A along dental outlines (C), along outlines of coronoid processes (D), along outlines of lower border of mandibles (E). Dark area, r, in E represents approximate extent of condylar resection. Note in the operated-side tracings the extensive deficiency of the posterior ramus and, in D and E, the teeth less fully erupted than on the unoperated side [from SARNAT and MUCHNIC, 1971a].

superposed on the teeth, and the differences of the levels of the zygomatic arches were demonstrated. When the tracings were superposed on the external auditory canals and the upper borders of the zygomatic arches, the lesser maxillary height of the operated right side was apparent (fig. 6D).

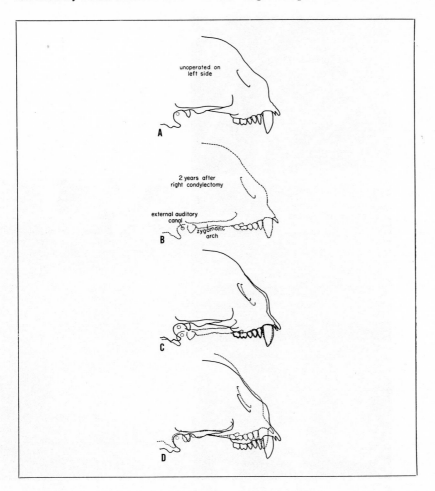

Fig. 6. Tracings of upper face from right and left lateral photographs of same adult squirrel monkey (No. 3–4) as in figure 5. *A* Unoperated left side (reversed to facilitate comparison). *B* Operated right side. *C, D* broken line of B superposed on solide line of A along dental outlines (C) and at external auditory canals and upper borders of zygomatic arches (D) [from SARNAT and MUCHNIC, 1971a].

B. Discussion

The condyle is capped by a narrow layer of avascular fibrous tissue that contains connective tissue cells and a few cartilage cells (fig. 7A, B). Under this fibrous covering is a chondrogenic layer that gives rise to

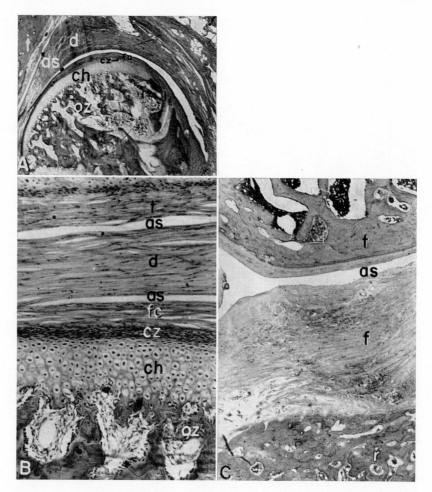

Fig. 7. Photomicrographs of sections of left and right temporomandibular joint regions of growing rhesus macaque No. 2–5, stained with hematoxylin and eosin; right mandibular condyle resected at age 8 months; animal killed at age 39 months; t = temporal bone; as = articular space; d = disc; fc = fibrous covering of condyle; cz = chondrogenic zone; ch = proliferating cartilage; f = fibrous tissue of false joint; oz = bone of condyle; r = upper part of remaining ramus. (Original magnifications: A: ×20; B: ×120; C: ×20). *A, B* Unoperated left temporomandibular joint, apparently within normal limits histologically; active growth of cartilage and bone occurring in epiphyseal-like region of condyle. *C* Area from which right mandibular condyle was resected, resulting in functional false joint devoid of cartilaginous growth site; note absence of cz. There is some apposition of bone however, at junction of f, r; wide zone of fibrous tissue serves as substitute for removed condyle and neck [from SARNAT, 1957].

hyaline cartilage cells constituting the second zone. In the third zone, destruction of the cartilage and ossification around the cartilage scaffolding can be seen. The cartilage in the head of the mandible is not homologous to an epiphyseal cartilage because it is not interposed between two bony parts; nor is it homologous to an articular cartilage, because the free surface bounding the articular space is covered by fibrous tissue.

In the group of growing rhesus macaques, *post mortem* histologic study of the unoperated left temporomandibular joint of a monkey at the age of about 39 months revealed an essentially normal joint and normally growing condyle (fig. 7A, B). Study of the right side, from which the condyle had been removed 31 months before death, however, revealed between the operated surface of the mandible and the skull a region about 3 to 5 mm wide filled with fibrous tissue and articulating surfaces with no disc interposed (fig. 7C). There was no evidence of cartilage being replaced by bone in the condylar-like area, but instead only a layer of dense fibrous tissue that was attached to the ramus inferiorly and articulated with the temporal bone superiorly.

In the squirrel monkeys, examination of the adult right mandibular condyle, resected at the beginning of the surgical experiment, revealed an outer layer of dense fibrous connective tissue, a middle zone of hyaline cartilage, and an inner region of bone and marrow spaces (fig. 8A, C), but no cartilage or bone formation. Examination of the intact left condyle resected at death two years later showed similar findings (fig. 8B, D).

Since the facial skeletal changes are similar (but not the same) in both the adult and growing monkeys after condylectomy, it seems that removal of that growth site, as in the young macaques, is not the principal responsible factor. Rather, the changes are probably secondary to disruption of normal temporomandibular joint function, including elimination of sensory receptors. Loss of the anatomical integrity of the temporomandibular joint, loss of function of the lateral pterygoid muscle, altered function of the medial pterygoid, masseter, temporal, and suprahyoid muscles, and establishment of a false joint all served to modify the direction and amount of muscular pull accompanying the altered position and motion of the mandible. These changes in function could influence the remodeling of the ramus and adjacent bony structures [SARNAT and MUCHNIC, 1971b].

Growth of bone occurs at the superoposterior surfaces of the mandible [ROBINSON and SARNAT, 1955; ENLOW, 1968; MOSS and RANKOW, 1968; SICHER and DUBRUL, 1970]. Whether the mandibular condyle is a

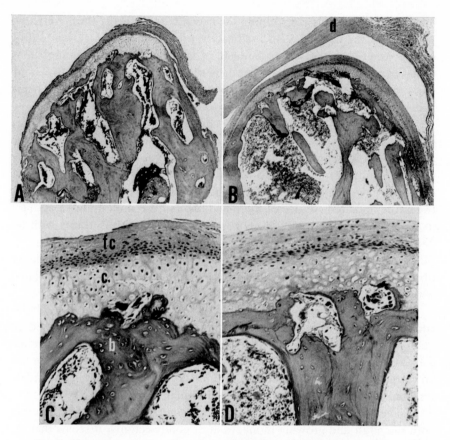

Fig. 8. Photomicrographs of sections of right (A, C) and left (B, D) mandibular condyles of adult squirrel monkey No. 3–8, stained with hematoxylin and eosin; right condyle resected at beginning of experiment; left one resected 2 years later at death. Note similar histologic characteristics in both condyles. Hyaline cartilage layer is inactive in process of osteogenesis, with no evidence of calcification, resorption, or replacement by bone (b); c = cartilage; d = disc; fc = articular fibrous connective tissue. (Original magnifications: A, B: ×45; C, D: ×200) [from SARNAT and MUCHNIC, 1971a].

primary or secondary site of growth is much discussed. Although the experiment in the adult monkey, as in the earlier experiments in the growing monkey, demonstrates the extreme loss of bone in the posterosuperior ramus two years after condylectomy, it does not solve that problem. The experiment does not prove that the condyle is not a growth site.

Summary and Conclusions

A series of surgical experiments carried out to evaluate normal and abnormal gross postnatal changes of the monkey facial skeleton were reviewed and summarized in relation to basic concepts of growth of bones. The method employed to study sutural growth activity involved the implantation of radiopaque silver amalgam markers in the bones combined with direct gross measurements and measurements taken from serial cephalometric radiographs. The increased separation of the implants (or implant images) on either side of a suture was taken as an indication of growth. Of five facial sutures studied in growing monkeys, growth was greatest at the zygomaticotemporal suture.

In another experiment, resection of the median and transverse palatine sutures in growing monkeys resulted in no gross facial deformity. The growth of bone which occurred at some of the facial sutures thus was probably mostly compensatory or secondary to some other factor. After surgical resection of the mandibular condyle in both growing and adult monkeys, however, severe deformities were obtained. The deficiency of ramus bone at *post mortem* almost three years later was more extensive than the amount resected. The important factor in this increased loss of bone was loss of integrity of the temporomandibular joint, rather than loss of a growth site.

Facial skeletal growth is a result of the synchronous coordination of the differential activities of the three types of bone growth and change at various growth sites. The dynamics of the growth and development of the facial skeletal system are a fascinating, complex, incomplete chapter of biology.

References

ENLOW, D. H.: The human face (Harper & Row, London 1968).

GANS, B. J. and SARNAT, B. G.: Sutural facial growth of the *Macaca rhesus* monkey: a gross and serial roentgenographic study by means of metallic implants. Amer. J. Orthod. *37:* 827–841 (1951).

MOSS, M. L. and RANKOW, R. M.: The role of the functional matrix in mandibular growth. Angle Orthod. *38:* 95–103 (1968).

ROBINSON, I. B. and SARNAT, B. G.: Growth pattern of the pig mandible. A serial roentgenographic study using metallic implants. Amer. J. Anat. *96:* 37–64 (1955).

ROY, E. W. and SARNAT, B. G.: Growth in length of rabbit ribs at the costochondral junction. Surg. Gynec. Obstet. *103:* 481–486 (1956).

SARNAT, B. G.: Facial and neurocranial growth after removal of the mandibular condyle in the *Macaca rhesus* monkey. Amer. J. Surg. *94:* 19–30 (1957).

SARNAT, B. G.: Palatal and facial growth in *Macaca rhesus* monkeys with surgically produced palatal clefts. Plast. reconstr. Surg. *22:* 29–41 (1958).

SARNAT, B. G.: Growth of bones as revealed by implant markers in animals. Amer. J. phys. Anthrop. *29:* 255–286 (1968).

SARNAT, B. G.: The face and jaws after surgical experimentation with the septovo-meral region in growing and adult rabbits. Acta oto-laryng., Stockh. *268:* suppl., pp. 1–30 (1970).

SARNAT, B. G.: Clinical and experimental considerations in facial bone biology: growth, remodeling and repair. Amer. dent. Ass. *82:* 876–889 (1971a).

SARNAT, B. G.: Surgical experimentation and gross postnatal growth of the face and jaws. J. dent. Res. *50:* 1462–1476 (1971b).

SARNAT, B. G. and ENGEL, M. B.: A serial study of mandibular growth after remov-al of the condyle in the *Macaca rhesus* monkey. Plast. reconstr. Surg. *7:* 364–380 (1951).

SARNAT, B. G. and MUCHNIC, H.: Facial skeletal changes after mandibular condylec-tomy in the adult monkey. J. Anat., Lond. *108:* 323–338 (1971a).

SARNAT, B. G. and MUCHNIC, H.: Facial skeletal changes after mandibular condylec-tomy in growing and adult monkeys. Amer. J. Orthod. *60:* 33–45 (1971b).

SCOTT, J. H.: Dento-facial development and growth (Pergamon Press, London 1967).

SICHER, H. and DUBRUL, E.: Oral anatomy; 5th ed. (Mosby, St. Louis 1970).

WEINMANN, J. P. and SICHER, H.: Bone and bones; 2nd ed. (Mosby, St. Louis 1955).

Author's address: Dr. BERNARD G. SARNAT, 435 North Roxbury Drive, *Beverly Hills, CA 90210* (USA)

Symp. IVth Int. Congr. Primat., vol. 3: Craniofacial Biology of Primates, pp. 227–240 (Karger, Basel 1973)

A Longitudinal Study of Cephalofacial Growth in *Papio cynocephalus* and *Macaca nemestrina* from Three Months to Three Years[1]

D. R. SWINDLER, J. E. SIRIANNI and L. H. TARRANT

Department of Anthropology and The Regional Primate Research Center, University of Washington, Seattle

Introduction

The investigation of craniofacial growth in nonhuman primates using cephalometric techniques has, unfortunately, lagged behind such studies in man. When these methods have been employed they have usually been experimental [GANS and SARNAT, 1951], cross-sectional [KROGMAN, 1931] or, at the very best, short-term serial studies [MICHEJDA and LA-MEY, 1971; ELGOYHEN *et al.*, 1972]. Frequently these studies have also had to rely upon animals of unknown chronologic age, since until relatively recently most monkeys used were born in the wild. Such prohibitions have been particularly constraining since it is well known that the efficacy of cephalometric analysis resides, in great part, in the serial or longitudinal study which permits the examination of the same individual (monkey) at selected intervals [SCOTT, 1967]. Since growth is change through time, it is important to measure the same animal more than once. Serial cephalometry allows this and thus provides data on the pattern and magnitude of craniofacial growth and permits comparative analyses, provided that one remembers 'that any differences observed are relative only to the common point or plane from which such differences are noted' [MOORE, 1971, p. 777].

The present study is concerned with the growth of the craniofacial complex in two genera of nonhuman primates, *Macaca* and *Papio. Maca-*

1 This research was supported in part by National Institutes of Health grants DE-02918 and RR-00166.

ca is the smaller animal. At birth the neonatal macaque weighs approximately 460 to 470 g [DILLINGHAM, 1972]; while its baboon counterpart averages 775 g [SNOW, 1967]. Also, at any time during the entire postnatal growth period the baboon is larger than the macaque, although the differences are less between the females of the two genera. The duration of growth time is also different in that the macaque reaches its maximum size at about seven years of age [GAVAN and SWINDLER, 1966]; while the evidence we have at present indicates a slightly longer period of growth for the baboon, perhaps as long as eight or nine years [GEAR, 1926; FREEDMAN, 1957]. Specifically then, we shall address ourselves to two problems: one, the kinds and amounts of linear and angular dimensional differences between these two genera during this early postnatal growth period; and two, an analysis of the relative growth rates of these dimensions in the two genera and an assessment of their significance.

Methods and Materials

The monkeys were born at the Regional Primate Research Center Field Station at Medical Lake, Washington. This colony was inaugurated several years ago to investigate the details of nonhuman primate growth and development. The express purpose of the colony was to study cephalofacial growth longitudinally in two samples of terrestrial quadrupedal monkeys, the pigtail monkey, *Macaca nemestrina*, and the olive baboon, *Papio cynocephalus*.[2] The parental population was free-ranging before capture and removal to the Primate Center.

The sample consists of 38 unrelated *Macaca nemestrina* and 16 *Papio cynocephalus*. The monkeys were weaned at approximately three months and thereafter raised in bisexual groups. A diet consisting mainly of monkey chow with daily additions of green vegetables plus proper medical attention has maintained excellent health in the colony. Of course, mild respiratory and enteric disorders are not uncommon among laboratory primates; and the present colony is no exception. There have been, however, no long-term debilitating illnesses which might affect the monkeys' growth and development.

The animals were anesthetized with Sernylan, then measured and X-rayed every three months for the first three years of life and biannually for the remainder of the growth period. Roentgenograms of the head in *norma lateralis* were obtained with the monkey fixed in a cephalostat that was designed specifically for primate

2   Recent studies of baboon taxonomy suggest that all savanna baboons represent one polytypic species, i.e. a single gene pool [BUETTNER-JANUSCH, 1966; THORINGTON and GROVES, 1970; MAPLES, 1972]. If this is correct, then the legitimate name for the olive baboon is *Papio cynocephalus* (Linnaeus, 1766) since this is the senior species. The future use of *anubis* and *doguera* for the specific name of the olive baboon should therefore be suppresed.

research. Kodak Industrial Type AA (ready pack) film was used during the study. All radiograms were taken at 1 sec, 85 k.v.p., and 300 m.a. This technique has been satisfactory for the younger monkeys (3 months to 3 years), but as they grow older we anticipate an increase in the k.v.p. All of the cephalometric films were taken with a subject-film distance of 12 cm and target-subject distance of 132 cm. Tracings of each cephalogram were made on 0.003 acetate paper by only one of us (J. E. S.). This minimized the possibility of introducing additional personal error into the tracing. In order to obtain the absolute values of the linear measurements taken from the tracings, each measurement was corrected for enlargement using the slide rule method devised by THUROW [1951].

The cranial, facial, and dental features were drawn and the following cephalometric landmarks were used: Nasion, sella turcica, prosthion, menton, basion, posterior nasal spine, articulare (Bjork), gonion, bregma, and interdentale inferior. Linear and angular measurements were used, and all dimensions were taken to the nearest 0.1 mm (fig. 1). These measurements were compared at each age level for sexual dimorphism within a species as well as for intergeneric differences. The significance of the difference was determined using a t-test. Additionally, several graphs were prepared in which a linear dimension was plotted against time (age in thousandths of a year). This permitted an assessment of the amount and tempo of absolute growth within and between the taxa.

The relative growth rates used in this paper were calculated using the method of FISHER [1921]. This permits an analysis of growth rate and growth time, which are important when one compares primate species of different adult size; for, as SCHULTZ noted some years ago, it is fundamental to determine 'whether body size is dependent upon the duration of growth or upon the speed of growth' [1956, p. 889]. The relative growth rates were calculated by dividing the differences in the natural logarithms of size by the differences in the time at which these two dimensions were obtained, thus:

$$\frac{\ln S_2 - \ln S_1}{T_2 - T_1},$$

where 1n S is the natural logarithm of size (measurement) and T is time. This calculation was done for each pair of adjacent measurements taken on an animal through time. Each relative growth rate was then placed in its appropriate three-month age interval, determined by the average age of the animal between the two adjacent times at which it was examined, i.e. $(T_2 - T_1)/2$. The relative growth rates of the cephalofacial dimensions were then compared by sex within each taxa and between the genera for each age range, and the significance of the differences was ascertained using a t-test.

Results and Discussion

The absolute linear dimensions for each age period were calculated, and the sexes were combined since there were very few significant differ-

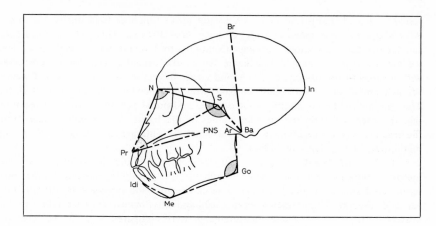

Fig. 1. Cephalometric landmarks, dimensions, and angles.

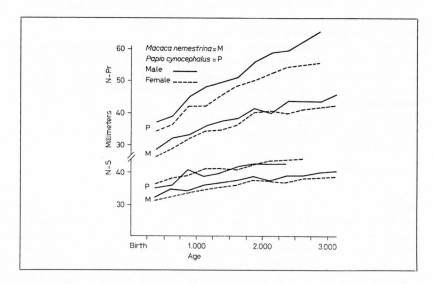

Fig. 2. Average nasion-sella and nasion-prosthion lengths by chronological age.

ences within the genera. We hasten to add, however, that the males were almost always larger than the females. The differences were usually slight, as can be observed in figures 2 and 3. On the rare occasion when the t-test revealed a statistically significant difference between the sexes, the dimension involved was one of the upper face. This difference is particu-

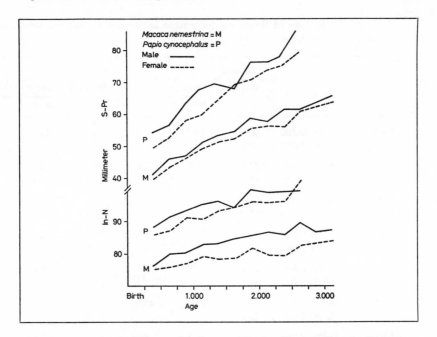

Fig. 3. Average inion-nasion and sella-prosthion lengths by chronological age.

larly noticeable in the nasion-prosthion and sella-prosthion measurements (fig. 2, 3). Also, it is greater within *Papio* than *Macaca* and thus may indicate the early domination of the downward growth of the face associated with a rotation of the facial axis on the basicranial plane, a well-known and important component of facial growth in the baboon [Zuckerman, 1926] that we suggest manifests itself at an earlier chronologic age in males than in females.

Whether these growth processes (presence or absence of sexual dimorphism) are related to menarche in these monkeys is difficult to determine. The evidence we have suggests that *M. nemestrina* achieves menkeys (1¹/₂–2¹/₂ yr) van Wagenen [1952]. According to Napier and Napier [1967], menarche in the baboon is somewhat later (3¹/₂–4 yr). The data at hand do not indicate a significant dimensional separation occurring between the sexes during these time periods, although the baboons are a little young for a positive statement (fig. 2, 3). This general lack of significant sex differences during the early growth period of cercopithe-

Table I. Comparison of three-year percentage increments of *Macaca nemestrina* and *Papio cynocephalus*

	Age	Macaca			Papio		
		Av, mm	Increment		Av, mm	Increment	
			mm	%		mm	%
N-Pr	0.250–0.500	27.2	16.2	60	35.7	23.3	65
	2.750–3.000	43.4			59.0		
N-S	0.250–0.500	31.8	8.3	26	35.6	10.6	30
	2.750–3.000	40.1			46.2		
N-In	0.250–0.500	76.2	11.1	15	87.3	15.0	17
	2.750–3.000	87.3			102.3		
S-Pr	0.250–0.500	40.4	23.0	57	51.6	31.5	61
	2.750–3.000	63.4			83.1		
PNS-Pr	0.250–0.500	25.6	17.7	69	31.5	22.5	71
	2.750–3.000	43.3			54.0		
Ar-Go	0.250–0.500	10.1	15.7	155	12.8	21.4	167
	2.750–3.000	25.8			34.2		
Me-Idi	0.250–0.500	13.1	11.3	86	14.7	13.2	90
	2.750–3.000	24.4			27.9		
Go-Me	0.250–0.500	26.5	18.2	69	35.4	21.2	60
	2.750–3.000	44.7			56.6		

coids has been noted earlier by GAVAN and SWINDLER [1966] and SNOW [1967].

After grouping the sexes, a t-test was calculated for each age level; and significant intergeneric differences were found for each age category. Thus, *P. cynocephalus* is larger in absolute size than *M. nemestrina* during the growth period studied. The growth increments from three months to three years, as well as the percentage increments for eight measurements are tabulated in table I.

The greatest change in both species involves the upper face (S-Pr and N-Pr); while the anterior limb of the cranial base (N-S) displays little growth. In a recent study of the flexion of the cranial base in *M. mulatta*, MICHEJDA and LAMEY [1971] noted a slow but gradual increase in the anterior cranial base. The mean increase for a group of juvenile animals (age 76–132 weeks) observed for 54 weeks was 3.25 mm. This portion of the cranial base seems, therefore, to be relatively inert for at least the first three years of postnatal life in these three cercopithecoids. Similarly, the cranium shows a small amount of increase (N to In) when compared with the face. This is particularly striking when the percentage increments are compared in table I. These figures express increments as percentages of growth already achieved, i.e. gain expressed as a function of its own progress [KROGMAN, 1950]. In the present analysis, the upper face (N-Pr) of the baboon shows a percentage increment of 65, compared with only 17⁰/o for cranial length (N-In). Occurring *pari passu* with the elongation of the upper face has been a downward displacement of prosthion (S-Pr) as registered from the cranial base plane. The similarity in the velocity of growth as defined by the percentage increment in these two linear dimensions is not surprising since there is a 0.94 correlation between them. These data augur well for the notable discrepancy between cranial and facial growth in cercopithecoids, particularly manifest in adult baboons.

Mandibular growth in both animals produces dramatic changes in shape during this time period. The height of the mandible (Ar-Go) more than doubles in size; while lower face height (Me-Idi) shows a high percentage increment of 86 for the macaque and 90 for the baboon. During this period of growth, mandibular length (Go-Me) shows about the same amount of increase as does the palatal length (PNS-Pr). Children with the equivalent dental ages of the monkeys in this study show percentage increments of 57 and 35 for mandibular height and length, respectively [KROGMAN and SASSOUNI, 1957]. Thus, orthognathous man shows about one-half of the percentage increment as these monkeys for the same dental ages.

Two angular measurements are presented in figure 4. The cranial base angle Ba-S-N shows a gradual increase in both animals. The opening up of the cranial base during early postnatal ontogeny is a slow process in these monkeys, but it is an important factor since it is primarily responsible for bimaxillary prognathism. In man, there is little increase in this angle after birth; and this small amount may simply be the result of an upward movement of nasion from the growth of the nasal on the frontal bone [SCOTT, 1967]. In their study of the cranial base of *M. mulatta*, MICHEJDA and LAMEY [1971] demonstrated a 'considerable increase' of the angle in the older juvenile monkeys that they attributed to the eruption of the second permanent molars. The older monkeys in the present sample are in the process of second molar eruption, and there is no apparent tempo change demonstrable. Moreover, although no data are given here,

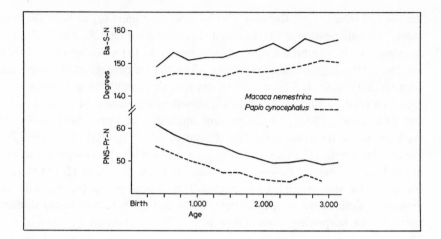

Fig. 4. Average angle changes plotted against chronological age.

we have a few monkeys over four years old (the sample size is too small to include in the present study) who similarly exhibited a gradual increase in this angle.

The baboons display a consistently smaller cranial base angle than the pigtails through the period studied. The approximate difference is four to six degrees for each age category. This would appear to be even more surprising since the baboon is considered the most prognathic of living primates. The answer, at least in part, resides in the fact that the baboon also presents the greatest downward displacement of the upper face, resulting in what Drennan called 'subgnathism' [quoted in ZUCKERMAN, 1926]. Thus, subgnathism refers to the downward, and prognathism to the forward, component of facial growth. The baboon begins and maintains a more acute cranial base angle than the pigtail, resulting in the elongated face so characteristic of adult baboons.

The other angles are more difficult to interpret. The increase of the nasion angle and the decrease of the angle at prosthion seem to be related to the various osseous adjustments going on during this ontogenetic period and, again, reflect the downward and forward growth of the muzzle. The vicissitudes in the gonial angle reflect the marked variability of the region in both monkey and man [ELGOYHEN *et al.*, 1972; KROGMAN and SASSOUNI, 1957]. The general decrease in this angle during ontogeny characterizes both man and monkey.

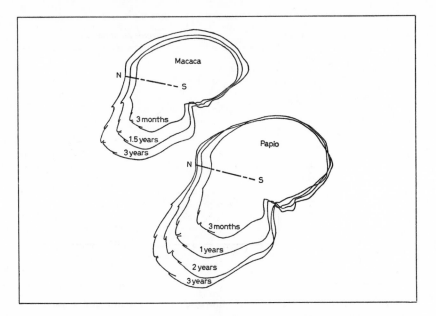

Fig. 5. Profile views of *Macaca* and *Papio* superposed on the nasion-sella plane.

To gain a better visual impression of the direction of craniofacial growth during the period studied, composite tracings were made of the baboon and pigtail skulls at three months and at one, two, and three years (fig. 5). The profiles are oriented on the Frankfort horizontal, and sella-nasion serves as the line of superposition.

The most outstanding feature revealed by these profile drawings is the change which takes place in the relative proportions of the cranial and facial skeletons. At each age, the discrepancy between face and cranium increases, culminating eventually with the morphological disparity exhibited by adults. During the period studied, the ratio of the growth of these two regions, as represented by sella to prosthion or nasion to prosthion on the one hand and on the other by basion to bregma or nasion to inion, was of the order of three to one. This relationship is true for both species, but the ratio is slightly higher for baboons. There is, therefore, a gradual and consistent relative diminution in the neurocranium and an increase in the prolongation of the muzzle with advancing age.

Of the relative growth rates studied, only seven were significant; and none displayed any sort of age-related pattern. That relative growth rates do not differ significantly during the first three postnatal years in these

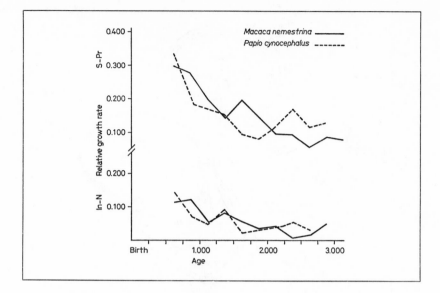

Fig. 6. Generic comparison of inion-nasion and sella-prosthion relative growth rates by chronological age.

two genera was recently demonstrated by SWINDLER and SIRIANNI [1972] for palatal growth. Moreover, GAVAN and SWINDLER [1966] found few significant differences in sitting-height growth rates between rhesus monkeys and chimpanzees after about one and one-half years. Prior to this time the macaque exhibited an appreciably faster rate than the chimpanzee, thus demonstrating a higher initial postnatal growth rate. In the present comparison, the relative growth rates were, in general, quite similar in tempo for the two species for each age catagory studied. Figures 6 to 8 support this statement, as well as display the slight vicissitudes in the rates between pigtails and baboons. In addition, they depict the decelerating nature of these growth rates through time.

The comparative rate of displacement of the upper facial skeleton (S-Pr) relative to the cranial base is shown in figure 6. The forward and downward rate of growth of the muzzle progresses at approximately the same rate in *Papio* and *Macaca*. The initial high postnatal rates decelerate steadily until about two years of age, when there is a slight rate increase in *Papio*. In the pigtail, the rates continue to decrease after the age of two and begin to plateau around two and one-half years, particularly in the sella-prosthion dimension. The rate of compensatory displacement of the

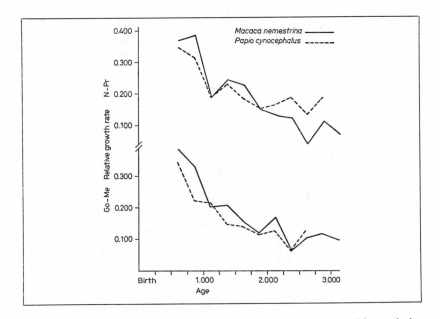

Fig. 7. Generic comparison of gonion-menton and nasion-prosthion relative growth rates by chronological age.

mandible (Go-Me) assists in maintaining the changing maxillo-mandibular relations during this time period (fig. 7). This observation is further substantiated by the high correlation coefficient between these dimensions, 0.95 (it should be noted that all of the animals had a class I molar relation during this interval). The similarity of these relative growth rates through time is an important factor in the harmonious development of the craniofacial complex of the monkey. The direction and magnitude of absolute change in this complex, relative to the cranial base, can be visually appreciated in figure 5.

Two measures of relative cranial growth rates are presented in figures 6 and 8 (In-N and Ba-Br, respectively). Here is the other aspect of the growth-rate differential presented by these monkeys. The rates of cranial growth are significantly slower when compared with facial growth. There is the same general deceleration as shown by the other dimensions, but the amount of rate change is much different; and, in addition, the earlier rates are much below those for the facial measurements. The differences between the facial and cranial growth rates demonstrate the degree of allometric growth occurring between these two regions dur-

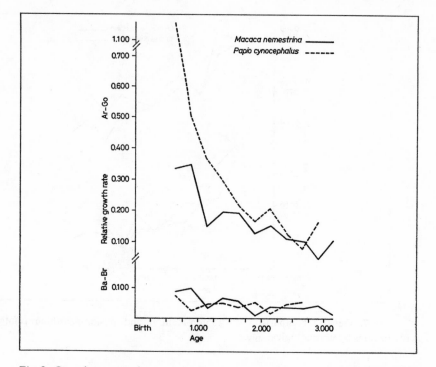

Fig. 8. Generic comparison of basion-bregma and articulare-gonion relative growth rates by chronological age.

ing the period studied. The facial growth rate pattern does not reach the slow velocity of cranial growth until the animals are approximately two and one-half years old. The rates of facial growth in the pigtail appear to decelerate slightly faster than they do in the baboon. These differential rates help to explain the gradual dominance of the facial skeleton over the cranium as age advances (fig. 5).

The pattern which emerges from the present analysis of these relative growth rates is one of gradual but constant deceleration and similarity of rates between the species. The baboons are larger animals at birth, and as adults they are larger than the pigtails. This larger size is maintained through their growth period by the consubstantiality between their growth rates. In addition, the size discrepancy between adult animals is probably also a function of growth time, in view of the fact that these similar rates proceed for different periods of time. The evidence to date would seem to suggest that relative growth rates among primates are constant and growth time is variable [FREEDMAN, 1962; GAVAN and SWINDLER, 1966; SWINDLER and SIRIANNI, 1972].

Summary

Cephalofacial growth was studied in *Papio* and *Macaca* for the first three years of postnatal life. There were no significant sex differences in the dimensions studied, but all diameters were significantly different between the genera. Both *Papio* and *Macaca* demonstrated similar amounts of allometric growth in their cephalofacial complexes during this time interval. Muzzle elongation was an important factor in both monkeys. There was no intrageneric sexual dimorphism in relative growth rates and only an occasional intergeneric difference. The baboon is the larger monkey, but its relative growth rates were similar to the pigtails; and in both genera the rates displayed a decelerating curve. Thus, both monkeys grow at approximately the same relative rates; and we suggest that the obvious size difference between the genera is genetically determined throughout growth and is not caused by a faster rate of growth for baboons, as might be supposed.

References

BUETTNER-JANUSCH, J.: A problem in evolutionary systematics; nomenclature and classification of baboons, genus *Papio*. Folia primat. *4:* 288–308 (1966).

DILLINGHAM, L. A.: Personal communication (1972).

ELGOYHEN, J. C.; RIOLO, M. L.; GRABER, L. W.; MOYERS, R. E. and McNAMARA, J. A., jr.: Craniofacial growth in juvenile *Macaca mulatta:* A cephalometric study. Amer. J. phys. Anthrop. *36:* 369–376 (1972).

FISHER, R. A.: Some remarks on the method formulated in a recent article on the quantitative analysis of plant growth. Ann. appl. Biol. *7:* 367–372 (1921).

FREEDMAN, L.: The fossil Cercopithecoidea of South Africa. Ann. Transv. Mus. *23:* 121–262 (1957).

FREEDMAN, L.: Growth of muzzle length relative to calvaria length in *Papio*. Growth *26:* 117–128 (1962).

GANS, B. J. and SARNAT, B. G.: Sutural facial growth of the *Macaca rhesus* monkey: a gross and serial roentgenographic study by means of metallic implants. Amer. J. Orthod. *37:* 827–841 (1951).

GAVAN, J. A. and SWINDLER, D. R.: Growth rates and phylogeny in primates. Amer. J. phys. Anthrop. *24:* 181–190 (1966).

GEAR, J. H. S.: A preliminary account of the baboon remains from Taungs, So. afr. J. Sci. *23:* 731–747 (1926).

KROGMAN, W. M.: Growth changes in the skull and face of the chimpanzee. Amer. J. Anat. *47:* 343–365 (1931).

KROGMAN, W. M.: A handbook of the measurement and interpretation of height and weight in the growing child. Monogr. Soc. Res. Child Devel. *13:* iii–68 (1950).

KROGMAN, W. M. and SASSOUNI, V.: A syllabus in roentgenographic cephalometry (Univ. of Pennsylvania, Philadelphia 1957).

MAPLES, W. R.: Systematic reconsideration and a revision of the nomenclature of Kenya baboons. Amer. J. phys. Anthrop. 36: 9–19 (1972).

MICHEJDA, M. and LAMEY, D.: Flexion and metric age changes of the cranial base in the Macaca mulatta. I. Infant and juveniles. Folia primat. 14: 84–94 (1971).

MOORE, A. W.: Cephalometrics as a diagnostic tool. J. amer. dent. Ass. 82: 775–781 (1971).

NAPIER, J. and NAPIER, P. H.: Handbook of living primates (Academic Press, London 1967).

SCHULTZ, A. H.: Postembryonic age changes; in HOFER, SCHULTZ and STARCK Primatologia. Handbook of primatology, vol. 1, pp. 887–964 (Karger, Basel 1956).

SCOTT, J. H.: Dento-facial development and growth, vol. 6 (Pergamon Press, New York 1967).

SNOW, C. C.: Some observations on the growth and development of the baboon; in VAGTBORG The baboon in medical research, vol. 1, pp. 187–199 (University of Texas Press, Austin 1967).

SWINDLER, D. R. and SIRIANNI, J. E.: Palatal growth rates in Macaca nemestrina and Papio cynocephalus. Amer. J. phys. Anthrop. 38: 83–91 (1973).

THORINGTON, R. W. and GROVES, C. P.: An annoted classification of the Cercopithecoidea; in NAPIER and NAPIER Old World monkeys: evolution, systematics and behavior, pp. 629–647 (Academic Press, New York 1970).

THUROW, R. C.: Cephalometric methods in research and private practice. Angle Orthod. 21: 104–116 (1951).

VAN WAGENEN, G.: Age at menarche of the laboratory rhesus monkey. Anat. Rec. 112: 436 (1952).

ZUCKERMAN, S.: Growth changes in the skull of the baboon, Papio porcarius. Proc. zool. Soc., Lond. 55: 843–873 (1926).

Authors' addresses: Dr. D. R. SWINDLER and L. H. TARRANT, Room 5, Burke Memorial Washington State Museum, University of Washington, Seattle, WA 98195 (USA); J. E. SIRIANNI, Department of Anthropology, State University of New York at Buffalo, Buffalo, N.Y. (USA)

Symp. IVth Int. Congr. Primat., vol. 3: Craniofacial Biology of Primates, pp. 241–257 (Karger, Basel 1973)

Occlusofacial Morphological Integration

(Homo sapiens, Alouatta caraya, Cebus capucinus)[1]

M. R. ZINGESER

Oregon Regional Primate Research Center, Beaverton

Introduction

Morphological integration denotes a complex of structurally interdependent components whose integrated totality is necessitated by functional requirements. Mammalian dentitions are eminently suitable models for quantitative studies that focus upon the co-relationships of components contributing to an integrated whole. Typically, mammalian dentitions are characterized by a precision of fit between apposing tooth crowns and crown parts that largely determines their functional efficiency. This high degree of morphological integration is the result of rigorous selection commensurate with the importance to survival of feeding and defense. The basic Therian occlusal pattern of precisely fitting homologous crown parts can be traced beyond the Cretaceous and is discernible in most orders, including the Primates [GREGORY, 1922; SIMPSON, 1936; PATTERSON, 1956; VANDEBROEK, 1961; CROMPTON, 1971; HERSHKOVITZ, 1971].

Occlusal patterns are species-specific, and occlusal fit is maintained within the framework of individual variations by controlling and integrating genetic factors [PATTERSON, 1956; BUTLER, 1961]. Morphological integrative traits appear early in odontogenesis and are evident in the coordinated growth and development of the crowns of occluding teeth [GAUNT, 1963; MARSHALL and BUTLER, 1966]. Furthermore, the integrative nature of the dentition appears to encompass the total crown-root complex [ZINGESER, 1968a, b].

1   Publication No. 655 of the Oregon Regional Primate Research Center, supported in part by grant No. RR00163-13 of the National Institutes of Health.

The Pearsonian coefficient of correlation (r) and its derivatives are commonly used to determine the degree of size correspondence between apposing, functionally complementary crowns and crown parts. These procedures, pioneered by OLSON and MILLER [1958] working with the dentition of *Aotes*, have since been applied to human material by MOO-REES and REED [1964] and GARN *et al.* [1965] among others. Their results quantitatively confirm the high degree of tooth crown morphological integration grossly evident in occlusal relationships.

Although occlusal fit is ultimately made possible by the aforementioned integrative size and form traits of the dentition, the tooth-to-tooth spatial orientation necessary to effect occlusal precision depends upon the proper growth, form, and function of numerous non-dental elements of the masticatory system and associated craniofacial complex. The integrated nature of the total system is brought into sharp focus when one considers the relationship between occlusal precision and characteristics of the tooth-alveolar bone interaction. Teeth will move through root-enveloping alveolar bone under the stimulus of very slight pressure imbalances, a fact well known to orthodontists. It follows that the specific spatial arrangement of teeth and tooth parts that comprises occlusal precision (i.e. normal occlusion) is one of dynamic equilibrium brought about and main-

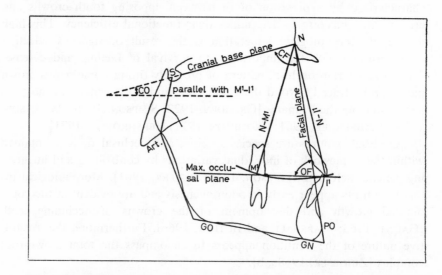

Fig. 1. Headplate tracings of a 12-year-old occlusal-normal girl. Compare the planes and angles with those in figure 2. Note the near-equality of the distances N–M¹ and N–I¹ and the arc construction of radius N–M¹.

tained by the form and functioning of a host of interacting, homeostatical-ly-regulated skeletal, neuromuscular, and oral environment structures.

I have hypothesized that the position of dynamic equilibrium of the dentition that constitutes normal occlusion with its attribute of species-wide relative pattern invariance should bear a dimensionally constant re-lationship within the facial skeleton that supports both the dentition and its associated position-influencing viscera. To assume the contrary would require of each individual an extraordinarily adaptive flexibility to ac-commodate the development and maintenance of the common occlusal pattern. In accordance with this reasoning, my objective has been to search for relatively invariant spatial relationships of the occlusal region within the facial skeleton.

My hypothesis has been substantiated by both X-ray cephalometric techniques (human clinical material) and direct craniometric data (*Al-ouatta caraya* and *Cebus capucinus*). The narrow range of variability that characterizes the orientation of the occlusal aspect of the maxillary denti-tion within the facial complex is quantitative proof of the high order of in-tegration that unites dentition and face into a form-functional entity. The purpose of this paper is to review the evidence and place the findings and associated conclusions in an appropriate bioclinical frame of reference.

Measuring Techniques

The orientation of the occlusal region is most readily quantified when it is re-duced to a series of consistently chosen, easily identifiable reference points. Theo-retically, an infinite series of such points is needed to define the total occlusal con-tour as these relate to skeletofacial structures. Practically, however, a pair of points suffices to define the orientation of the segment of occlusion they encompass. It was deemed sufficient to test the orientation properties of the maxillary occlusal region, since the normally close apposition of teeth makes the separate determination of mandibular occlusal orientation redundant.

Two methods are used to analyze the orientation properties of the occlusal re-gion:

(1) The first technique is applicable only to lateral X-ray cephalometric trac-ings. In application, a straight line joins a pair of molar and incisor occlusal surface reference points, usually M^1 and I^1 (fig. 1). The line represents an approximation of the position in space of the occlusal region included between M^1 and I^1. This is termed the maxillary occlusal plane.[2] This plane, based upon actual occlusal surface

2   Cephalometric and craniometric reference points and planes used in my inves-tigations are defined in detail in ZINGESER [1960, 1966].

loci, must be differentiated from the commonly used cephalometric occlusal plane, which is a construct arrived at by bisecting the incisal overbite. The slope or angular properties of the maxillary occlusal plane is a useful measure of orientation.

(2) The relative distances of occlusal surface reference points to skeletal reference points is another useful mensurational tool which lends itself to both roentgenographic and direct craniometric approaches. To illustrate the rationale underlying this technique, let us assume that the relative distances that the tips of all of the maxillary buccal cusps lie from nasion within a skull are determined. With these data, one can reconstruct a fairly accurate approximation of the buccal-occlusal contour within the skull. In practice, this is done by using one distance, say nasion-to-M^3, as a yardstick or basis for comparison. The corresponding cusp of M^2 might be 0.5 mm further from nasion, the corresponding cusp of M^1 1.2 mm closer to nasion, etc.

If these relative positions of tooth reference points as related to a skeletal reference point (nasion) show only slight variability in population samples, the following conclusions are justified:

(a) The occlusal contours do not vary greatly among individuals in the sample, and

(b) The occlusal aspect of the dentition relates to nasion in much the same manner in most members of the sample.

In summary, estimates of the parameters of variability of the relative distances of any pair of strategically located occlusal reference points to a specific skeletofacial reference point can be useful in assessing the contour and orientation stability of that portion of the occlusal region encompassed by these points.

X-Ray Cephalometry

The search for a common denominator of occlusofacial or occlusocranial orientation commensurate in its invariance with the well-documented phylogenetic constancy of occlusal relationships began with a study of clinical standardized cephalometric X-rays. Tracings of headplates of children and young adults with normal occlusions were superimposed upon each other in pairs along commonly used cranial base contours and planes. It soon became evident that the orientation of facial structures, including the dentition, to the neurocranial base normally encompasses a very wide range of variation (fig. 2).

In contrast to the wide range of variation in the orientation of splanchnocranial elements to the neurocranium, the orientation of maxillary occlusal reference points within the bony facial skeleton was found to accord with the hypothetically predicted constancy among both occlusal normal individuals and a variety of maloccluded types [ZINGESER, 1951, 1959, 1960]. This is graphically illustrated in figure 2, which represents

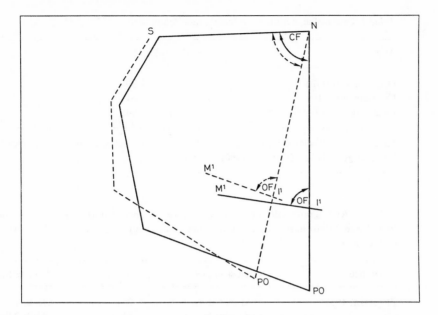

Fig. 2. Facial diagrams of two 12-year-old occlusal-normal boys superimposed upon the anterior cranial base plane N–S and registered at N. The nonconcordance of facial structures to cranial base is characteristic. In contrast, the intrafacial angle OF is identical in these two tracings [modified after Zingeser, 1960].

the superimposed facial diagrams of the two 12-year-old occlusal normal boys. The diagrams are superimposed upon the cranial base line nasion-sella and illustrate an extreme degree of nonconcordance of facial structure. However, when superimposed along the facial plane (nasion-pogonion), the diagrams show a considerable degree of concordance in the orientation of facial and intrafacial structures (especially the occlusal contact regions).

This contrast between intrafacial concordance and craniofacial non-concordance in cross-sectional material is brought home by differences in variability of the intrafacial versus craniofacial angles. Table I summarizes the statistical properties of the angular relationships of the maxillary occlusal region within the facial complex (occlusofacial angle OF, fig. 1 and 2) and to cranial base (cranio-occlusal angle CO, fig. 1 and 2). The statistical properties of the cranial base to the facial plane angle (angle CF) are also given. The cant of the maxillary occlusal plane within the bony face shows a high degree of orientation stability. This contrasts with

Table I. Occlusofacial and craniofacial angles compared

Angle	Mean ± SD
Occlusofacial (OF)	80.28° ± 1.89°
Craniofacial (CF)	79.71° ± 3.64°
Cranio-occlusal (CO)	19.61° ± 3.84°

Based upon cephalometric tracings of 42 12-year-old occlusal normal children (21 boys, 21 girls) [after ZINGESER, 1960] (see fig. 1 and 2).

Table II. Radiographically determined relative distances (mm) of paired occlusal loci I^1 and M^1 to nasion (i.e., the distance N to I^1 minus the distance N to M^1); clinical material

Reference	Occlusion	n	Mean ± SD
ZINGESER, 1951	mild crowding	25	0.76 ± 1.93
	retrognathic	51	1.33 ± 2.09
ZINGESER, 1960	normal	42	0.83 ± 1.12

its relationship and that of the facial plane to the cranial base (angles CO and CF). The slight variability of angular orientation of the occlusal region with the face is seen as evidence of a common pattern of adaptation of facial structures to the requirements imposed by a common pattern of occlusion. In contrast, no such significant consistency can be seen in the orientation of occlusal region to cranial base; nor is there any functional reason to expect consistency. The common practice of relating occlusal and facial components to reference sites of the cranial base in cross-sectional samples for the purpose of deriving clinically useful data originates with BROADBENT's [1937] application of neurocranial reference structures to longitudinal growth studies. BROADBENT's valid application of the early maturing, slow-growing cranial base region as a reference site from which to assess individual craniofacial growth loses its relevance when misapplied to cross-sectional material. When it is thus misapplied, the normally occurring wide range of variation in the orientation of the face to the neurocranium obscures an underlying and bioclinically significant occlusofacial pattern invariance [ZINGESER, 1959, 1960].

Table III. Relative distances[1] of paired occlusal loci to skeletofacial points

Skeleto-facial point	Paired occlusal loci	Relative distances[1], mm, mean ± SD		
		A. caraya (1)	*C. capucinus* (2)	*H. sapiens* (3)
N	I^1–M^1			0.83 ± 1.12
	I^1–M^3		–1.82 ± 1.28	
	I^1–M^2	0.48 ± 0.81		
	P^1–I^1		1.54 ± 0.78	
	P^1–M^3		–0.40 ± 0.76	
	P^1–M^2	–0.44 ± 0.66		
NLM	I^1–M^2	3.16 ± 0.98		
	P^1–M^2	–0.21 ± 0.69		
ZF	I^1–M^2	15.10 ± 1.02		
	P^1–M^2	5.35 ± 0.77		
ZM	I^1–M^3		8.17 ± 1.64	
	I^1–M^2	15.73 ± 1.18		
	P^1–M^3		–0.36 ± 1.07	
	P^1–M^2	4.00 ± 0.91		
	P^1–I^1		–8.44 ± 1.05	
APA	I^1–M^2	9.35 ± 1.61		
	P^1–M^2	–6.81 ± 1.03		

Skeletofacial reference points: N = nasion, NLM = naso-lachrymo-maxillary, ZF = zygomatico-frontal, APA = apex piriform aperture (see fig. 3). Data information: sample size (1) *A. caraya* 33–37 males [ZINGESER, 1966]; (2) *C. capucinus* 26–29 males; (3) *H. sapiens* 21 males and 21 females.

1 Relative distances are expressed as the mean differences between the distances of the first minus the second locus of each pair from the skeletofacial point indicated for each pair; for example: the distance N to I^1 minus the distance N to M^1, or the distance ZM to P^1 minus the distance ZM to M^3, etc.

Measuring Variability

As a measure of absolute variability, the standard deviation is used to assess data that derives from the application of the two measuring techniques mentioned above. Both occlusofacial and craniofacial angular measurements and measurements of the relative distances of occlusal surface reference points to skeletal reference points are concerned with describing morphological *relationships* rather than the *size* of structures.

Table IV. Upper anterior dentofacial height (mm), nasion to I[1]

Sample	Sex	n	Mean ± SD	V.
H. sapiens	M+F	42	79.85 ± 4.53	5.67%
A. caraya	M	24	38.89 ± 2.73	7.01%
C. capucinus	M	29	36.42 ± 2.07	5.71%

Table V. Mean ± SD (mm) of dental arch segment lengths in *Cebus capucinus*

Measurement	n=29 ♂	n=26 ♀	p
P[1]–I[1]	19.81 ± 0.88	17.95 ± 0.85	<0.001
P[1]–M[3]	18.63 ± 0.62	18.70 ± 0.66	N.S.

This distinction is relevant in the choice of statistical procedures. For example, the coefficient of variability (V) is widely used in biological studies. Its use should be confined to comparing homologous structures (elephant and shrew teeth) where presumably the size of the dispersion, and hence of the variability of the variate, is influenced by the order of magnitude of the structures being compared [SIMPSON *et al.*, 1960]. I apply the coefficient of variability in comparing species variability in dentofacial height (table IV).

The use of the coefficient of variability in comparisons involving the variability of morphological relationships can lead to completely erroneous conclusions. For example, the angular relationship between the anterior cranial base plane and the facial plane (angle CF, fig. 1 and 2 and table I) has a mean value of 79.71°. The angle of cranial base plane to occlusal plane (angle CO) has a mean value of 19.61°. The standard deviations, and hence the absolute variability of these angles, are comparable (3.64° and 3.84°). Relating the means of these angles to their standard deviations in deriving the coefficient of variability

$$\left[V = \frac{100 \times SD}{M} \right]$$

yields results that imply a considerably greater variability for angle CO as compared with angle CF; whereas the angular deviations of these structural relationships about their means are, in fact, of comparable magnitude.

Viewed in this context, the absolute angular variability of the occlusal region within the face (angle OF), with its standard deviation of 1.89°, markedly contrasts with the cranial base-to-facial structure angles CF and CO with their respective standard deviations of 3.64° and 3.84° (fig. 1 and 2; table I). The angular stability of the intrafacial orientation of the occlusal region is evidence of a high degree of morphological integration uniting the occlusal region within the facial complex.

The constituent reference points of the maxillary occlusal plane (M^1 and I^1) themselves relate to the nasofrontal suture (nasion = N) so that their relative distances to N show small absolute variability. As measured on headplate tracings, the distances $N–I^1$ and $N–M^1$ are nearly equal. If one uses the distance $N–M^1$ as a 'yardstick' as described in the Measuring Techiques section, $N–I^1$ averages 0.83 mm longer than $N–M^1$ with a standard deviation of 1.12 mm in occlusal normals. A variety of malocclusions were also measured for estimates of the parameters of these occlusofacial relationships. The results are given in table II.

The near equality of the distances of these occlusal reference points to nasion affords a rapid and convenient graphic device for clinically assessing occlusofacial relationships, their deviations, and the nature of corrective changes [ZINGESER, 1951, 1960, 1964, and in press]. An arc of radius $N–M^1$ (the 'yardstick') will intercept close to the tip of I^1 in normal occlusions and in a variety of malocclusions. Furthermore, the tips of the buccal cusps of the teeth intervening between I^1 and M^1 (not illustrated) will normally lie slightly above this arc (fig. 1).

I stress that the near equality of $N–M^1$ and $N–I^1$ seen in the lateral cephalic X-ray image has no functional significance *per se*. In fact, these distances are not equal craniometrically; they reflect a uniform and standardized distortion imposed by the X-ray technique. This element of standarized distortion does not detract from the validity of conclusions relating to the data thus derived. These data clearly indicate that the occlusal aspect of the maxillary dentition relates to the nasofrontal suture in a relatively invariant manner.

Although the statistical validation of these characteristics are based upon studies of Caucasians, the near equality of $N–I^1$ and $N–M^1$ seen in the X-ray image applies equally to other racial types and appears to be a species-wide attribute. I suspect that the cant of the maxillary occlusal plane to the facial plane may show some racial differences associated with varying degree of prognathism, but this characteristic remains to be investigated.

Age Changes in Human Beings

Age changes in these occlusofacial relationships were studied in two longitudinal samples of occlusal normal children [ZINGESER, 1960]. Changes are small and thus confirm the parallel nature of facial growth determined at the individual level by BROADBENT [1937] using cranial base reference planes. From age 8 to 12 years, $N–I^1$ shows an average increase of 1.12 ± 0.21 mm over $N–M^1$. From age 12 to 18 years, the growth of the nasion to molar component $N–M^1$ exceeds the growth of the nasion to incisor component $N–I^1$ by 0.12 ± 0.20 mm. These changes positively correlate with overbite changes with age and substantiate BRODIE's [1953] findings.

Craniometric Studies

Direct craniometric studies were undertaken to circumvent the limitations imposed by the X-ray image. Since human crania in sufficient numbers and in desirable condition were lacking, the skulls of two species of South American monkeys were studied (*A. caraya* and *C. capucinus*). Mature skulls were selected because the platyrrhine monkeys in common with most nonhuman primates and other mammals, and in contrast to man, develop a progressive occlusofacial prognathism with growth. For the sake of conciseness, only data from male skulls are given in this paper.

Specifically, the objectives were to broaden the base of information afforded by clinical cephalometric X-ray material by direct measurement testing of the variability of different paired occlusal surface loci to a variety of skeletofacial sites. In *A. caraya,* the black howler monkey, paired occlusal surface loci I^1M^2 and P^1M^2 were tested for variability of relative distances in relation to five skeletofacial sites [ZINGESER, 1966]. The occlusal surface loci are the tips of the paracones (mesiobuccal cusps of the molars; buccal cusps of the premolars). The skeletofacial reference points are defined as follows:

(1) Nasion (N): the mid-sagittal point on the nasofrontal suture.
(2) Naso-lacrimo-maxillary point (NLM): point of juncture of nasal and lachrimal bones with the maxilla; located close to the orbital rim.
(3) Zygomatico-maxillary point (ZM): point of juncture of the orbital rim with the zygomatico-maxillary suture.

(4) Zygomatico-frontal point (ZF): point of juncture of the orbital rim with the zygomatico-frontal suture.

(5) Apex of the piriform aperture (APA): highest point of the bony rim of the external nares; also the free edge of the internasal suture.

These reference points are figured in a drawing of a howler monkey skull in ZINGESER [1966] and are illustrated in a male *C. capucinus* skull (fig. 3), which shows additional points specifically relevant to craniometrics in that species. In the *Cebus* material, the relative distances and variabilities of distances of paired occlusal loci I^1M^3, P^1M^3, and P^1I^1 to nasion and to the zygomatico-maxillary point are determined. Cebid and human data are given in table III.

Topographic Variability and Relationships

Paired occlusal surface loci relate to the sutural sites with small absolute variability in all three species (table III). The significance of the size of the standard deviations associated with these occlusofacial topographic relationships is made more meaningful when compared with estimates of other occlusal and occlusofacial parameters. For example, the upper anterior dentofacial height ($N–I^1$) for the three species is given in table IV. The coefficients of variability (V) indicate that human and *Cebus* skulls are equally variable in this dimension and that the howler monkey is somewhat more variable. When tables III and IV are compared, it is evident that the upper facial orientation of the occlusal aspect of the maxillary dentition maintains its stability independent of considerable variations in upper dentofacial height.

Estimates of parameters of dental arch segment lengths of *C. capucinus* are given in table V. When compared with the *Cebus* data listed in table III, the order of magnitude of orientation variability is seen to approach the actual size variations of the lengths of the dental arch segments themselves.

The howler monkey data is the least variable of the three species samples studied. In the two cebid species, loci encompassing larger segments of the dentition are more variable (fig. 4). This reflects the increased variability expected with the combined dimensions of a greater number of teeth. In paired loci encompassing the maxillary canine tooth diastema, there is the added variability of this region.

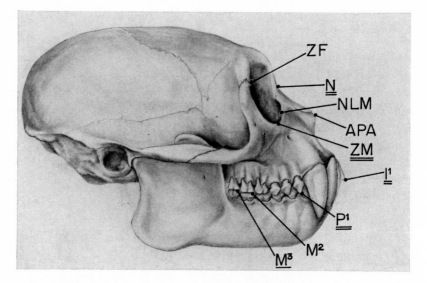

Fig. 3. Right lateral view of the skull of a male *Cebus capucinus.* Reference points underlined twice are shared in common with *Alouatta caraya* [Zingeser, 1966]. Those underlined once are unique to *Cebus,* and reference points that are not underlined refer only to the craniometric studies of *Alouatta.*

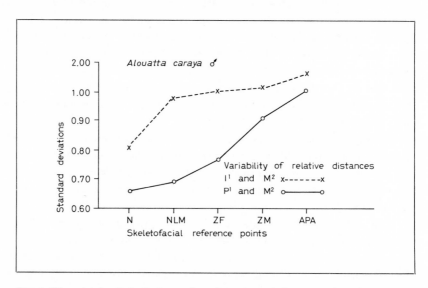

Fig. 4. The standard deviations plotted against their respective skeletofacial reference points for two groups of paired occlusal loci. Compare with table III.

The orientation of paired occlusal loci to nasion is the least variable finding (table III and fig. 4). This stability is associated with the midsagittal location of nasion. More lateral sites show larger standard deviations that reflect variability in skull width vis-à-vis maxillary dental arch width. However, another midsagittal point, the apex of the piriform aperature is the most variable in *A. caraya*. APA is the only skeletofacial point not completely surrounded by bone. Its free edge consists of thin, friable bone and this, rather than intrinsic characteristics of the site, is believed to account for its relatively large variability.

The sutural skeletofacial reference sites (fig. 3) are all associated with skeletal elements of cephalic capsules [Moss, this volume, pp. 191–208]. Nasion, the most stable in relationship to the occlusal aspect of the dentition, also occupies a uniquely stable position relative to the growing neurocephalic, nasal, and orbital capsules [Zingeser, 1966, fig. 4, p. 176]. It is situated at the juncture of these capsules. Three of the remaining four reference sites are on or close to the rim of the bony orbit. The fourth (apex of the piriform aperture) relates to the nasal capsule.

The precision with which these sutural sites are oriented to the occlusal (principal functioning) aspect of the dentition must have high adaptive significance [Muller, 1947/1948; Farris, 1966]. I postulate the following correlations in providing a working hypothesis.

(1) The facial orientation of the sutures is critical in guiding form, differential growth, and, by interaction with contiguous capsules, the growth vectors of their associated cephalofacial capsules. Supporting evidence is provided by Weinmann and Sicher [1947] who describe the direction of facial sutures as consistent with facial growth direction.

(2) The topographic relationships of sutures to the occlusal region reflect adaptations for the absorption and dispersion of occlusal stress in a manner consonant with the growth-accommodating functions of the sutures. This view compliments Moss' [1961] determination of extrinsic factors influencing sutural anatomy and Moffett's observations [this volume, pp. 180–190] on the effects of mechanotherapy.

Occlusofacial Integration

The integrated nature of the upper occlusofacial complex emerges when one considers the small variability of relationships of skeletal to occlusal sites in the context of the varied structures that lie between the points measured (fig. 3). These include the orbits, nasal cavity, maxillary sinuses, alveolar bone, and teeth.

I had proposed [ZINGESER, 1966] that this subcerebral, supra-occlusal region constitutes *in toto* a functional matrix, the function of which is to maintain a rigid dimensional constancy as a channeling device to facilitate adaptations for the vital processes of attaining and maintaining occlusal precision [cf. Moss and SALENTIJN, 1969a, b; and Moss, this volume, pp. 191–208]. This region should more properly be termed a functional cranial component. That this functional cranial component consists of subsets of functional matrices poses no conceptual problem. The structures interposed between the roof of the orbits and the occlusal region have many different functions which tend to shape their individual forms. No paradox is seen in conceiving that their combined forms function to stabilize the occlusal region in a relatively invariant pattern.

Comparative studies, especially where extreme specializations are present, are useful to elucidate form-functional relationships. Tarsius is a primitive nocturnal primate characterized by enormous eyes. In fact, the bony orbits dominate the skull to such an extent that the facial skeleton is reduced to almost avian proportions (fig. 5); and the olfactory region is severely limited. Obviously, the size and form of the optic capsules greatly influence total facial development and final configuration. One might just as easily have chosen an animal in which the nasal capsule is dominant. Either example serves to underscore the point I wish to make: the combined size and form of the upper face capsules determine the *basic* orientation of the maxillary tooth-bearing region.

The ultimate, precise orientation of the occlusal contact region is mediated largely by neuromuscular mechanisms that act upon teeth and their supporting pressure-labile alveolar bone. I visualize the stable upper occlusofacial functional component as a 'template' to which the mandible with its adaptive condyle and the mandibular dentition accommodate through the agency of the orofacial sensorium and associated muscles of mastication, lips, and tongue [KAWAMURA, 1963]. The resulting position of equilibrium at the occlusal interface reflects the integrated aggregate actions of the feed-back processes that regulate growth, form, and function. Homeostasis is the prime biological process [see LERNER, 1954], channeling form and function to the vital requirements of occlusal fit. The resulting commonality of occlusofacial pattern, mensurationally demonstrable at growth termination in the two cebid spesies and throughout growth in man, can aptly if incompletely be termed morphological integration.

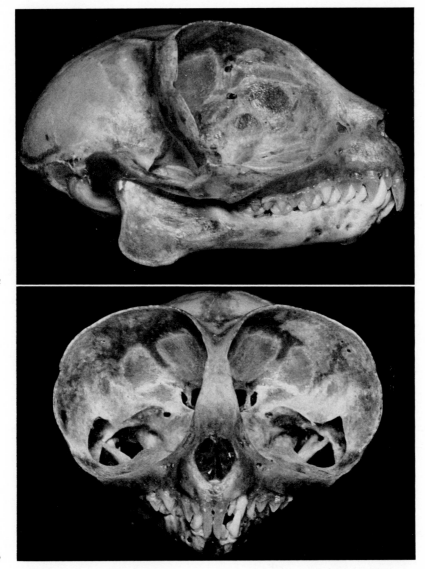

Fig. 5. Right lateral (A) and facial (B) views of the skull of a male *Tarsius syrichta*. The role of the facial capsules, in this case especially the orbital capsules, in determining facial configuration and, hence, the basic orientation of the maxillary dento-alveolar region is evident.

Acknowledgments

I appreciate the cooperation given by staff members of the Oregon Regional Primate Research Center in preparing the manuscript. I am particularly grateful to Mrs. BARBARA HOLLAND for her typing expertise, to Mr. JOEL ITO for his skillfully executed illustrations, and to Mr. HARRY WOHLSEIN for his photography. My thanks also go to Dr. C. O. HANDLEY of the US National Museum for his courtesy and cooperation in making available to me the Museum's *Cebus capucinus* skull collection.

References

BROADBENT, B. H.: The face of the normal child. Angle Orthod. *7:* 183–207 (1937).

BRODIE, A. G.: Late growth changes in the human face. Angle Orthod. *23:* 146–157 (1953).

BUTLER, P. M.: Relationships between upper and lower molar patterns. Int. Colloq. on the evolution of lower and non-specialized mammals. Kon. Vl. Acad. Wet. Lett. Sch. Kunst. België, Brussels; part I, pp. 117–126 (1961).

CROMPTON, A. W.: The origin of the tribosphenic molar in early mammals; in KERMACK and KERMACK Early mammals; suppl. No. I to Zool. J. Linnean Soc. (Zool.), vol. 50, pp. 65–87 (Academic Press, New York 1971).

FARRIS, J. S.: Estimation of conservation of characters by constancy within biological populations. Evolution *20:* 587–591 (1966).

GARN, S. M.; LEWIS, A. B., and KEREWSKY, R. S.: Size interrelationships of the mesial and distal teeth. J. Dent. Res. *44:* 350–354 (1965).

GAUNT, W. A.: An analysis of the growth of the cheek teeth of the mouse. Acta anat. *54:* 220–259 (1963).

GREGORY, W. K.: The origin and evolution of the human dentition (Williams & Wilkins, Baltimore 1922).

HERSHKOVITZ, P.: Basic crown patterns and cusp homologies of mammalian teeth; in DAHLBERG Dental morphology and evolution, pp. 95–150 (University of Chicago Press, Chicago 1971).

KAWAMURA, Y.: Recent concepts of the physiology of mastication; in Advances in oral biology, vol. 1, pp. 77–109 (Academic Press, New York 1963).

LERNER, I. M.: Genetic homeostasis, (Oliver & Boyd, Edinburgh 1954).

MARSHALL, P. M. and BUTLER, P. M.: Molar development in the bat *Hipposideros beatus* with reference to the ontogenetic basis of occlusion. Arch. oral Biol. *11:* 949–965 (1966).

MOOREES, C. F. A. and REED, R. B.: Correlations among crown diameters of human teeth. Arch. oral Biol. *9:* 685–697 (1964).

MOSS, M. L.: Extrinsic determination of sutural morphology in the rat calvaria. Acta anat. *44:* 263–272 (1961).

MOSS, M. L. and SALENTIJN, L.: The primary role of functional matrices in facial growth. Amer. J. Orthod. *55:* 566–577 (1969a).

Moss, M. L. and Salentijn, L.: The capsular matrix. Amer. J. Orthod. *56:* 474–490 (1969b).

Muller, H. J.: Evidence of precision of genetic adaptation; in Harvey Lectures, vol. 43 (Thomas, Springfield 1947/1948).

Olson, E. C. and Miller, R. L.: Morphological integration (University of Chicago Press, Chicago 1958).

Patterson, B.: Early Cretaceous mammals and the evolution of mammalian molar teeth. Fieldiana, Geol. *13:* 1 (Chicago Natural History Museum, 1956).

Simpson, G. G.: Studies of the earliest mammalian dentitions. Dent. Cosmos *78:* 791–800, 940–963 (1936).

Simpson, G. G.; Roe, A., and Lewontin, R. C.: Quantitative zoology; revised ed., p. 90 (Harcourt, Brace & Co., New York 1960).

Vandebroek, G.: The comparative anatomy of the teeth of lower and non-specialized mammals. Int. Colloq. on the evolution of lower and non-specialized mammals. Kon. Vl. Acad. Wetensch. Lett. Sch. Kunst. België; part I, pp. 215–320 (Brussels, 1961).

Weinmann, J. P. and Sicher, H.: Bone and bones. Fundamentals of bone biology; p. 93, fig. 58 (Mosby, St. Louis 1947).

Zingeser, M. R.: A linear property characteristic of the vertical orientation of certain maxillary dental units. Angle Orthod. *21:* 205–212 (1951).

Zingeser, M. R.: Dento-facial constancy unobscured by cranio-facial positional variance; Amer. Bd. of Orthod. Thesis (1959).

Zingeser, M. R.: Nasomaxillary proportional constancy. Amer. J. Orthod. *46:* 674–684 (1960).

Zingeser, M. R.: Vertical response to class II division I therapy. Angle Orthod. *34:* 58–64 (1964).

Zingeser, M. R.: Occlusofacial relationships in the mature howler monkey (*Alouatta caraya*). Amer. J. phys. Anthrop. *24:* 171–180 (1966).

Zingeser, M. R.: Functional and phylogenetic significance of integrated growth and form in occluding monkey canine teeth (*Alouatta caraya* and *Macaca mulatta*). Amer. J. phys. Anthrop. *28:* 263–270 (1968a).

Zingeser, M. R.: Sexual dimorphism in monkey canine teeth; in Proc. 8th int. Congr. Anthrop. Ethnol. Sci., vol. 1, pp. 305–309 (Science Council of Japan, Tokyo 1968b).

Zingeser, M. R.: Analyzing overbite problems and corrective changes through the application of intrafacial cephalometric guidelines (in press).

Author's address: Dr. Maurice R. Zingeser, Oregon Regional Primate Research Center, 505 N.W. 185th Avenue, *Beaverton, OR 97005* (USA)

Symp. IVth Int. Congr. Primat., vol. 3: Craniofacial Biology of Primates,
pp. 258–261 (Karger, Basel 1973)

Appendix

Nonhuman Primates as a Resource in Craniofacial Research[1]

K. K. HISAOKA

Deputy Associate Director for Extramural Programs, National Institute of
Dental Research, Bethesda, Maryland

History of NIDR

Twenty-four years ago, in 1948, Congress established the National Institute of
Dental Research (NIDR) assigning it the mission of improving the dental health of
the people of the United States through research. Among the oldest Institutes at the
National Institutes of Health (NIH), the NIDR will celebrate its silver anniversary
next year. Although the NIDR has supported a broad range of research in oral biology since the early 1950's, a significant increase of funds was not available until
1960.

In 1966, the Extramural Programs was organized into four categorical areas so
that NIDR could more effectively administer research and training programs designed to attack the major dental diseases and disorders; dental caries, periodontal
disease, oral-facial anomalies, and biomaterials. Today we have six program areas;
each supports research and training through grants and contracts. The research
grants include program projects that support multidisciplinary teams and Special
Dental Research Awards that provide small grants to newly-trained investigators. In
addition, each program area provides research Training Grants, Fellowships, and
Research Career Development Awards, which provide salaries for young scientists.

In 1971, with Congress and the nation addressing themselves to research results, the NIDR developed a targeted National Caries Program. Congress earmarked
$ 900,000 in new funds for the NCP, which involves NIDR intramural scientists
and experts in research institutions and in industry from all over the country. All
are seeking ways to make tooth decay largely preventable. Some investigators are
studying cariogenic microorganisms, others the effects of diet, and still others are
examining the host's resistance to tooth decay. In addition, preventive measures are
being tested; and new ones are being developed.

Research concerning another dental problem, periodontal disease, was increased
in fiscal 1972 when an additional $ 760,000 was allocated for investigations into

1   See editor's mention of Dr. HISAOKA's remarks in the Preface to this volume,
p. VIII. [Ed.]

the etiology, pathogenesis, and treatment of periodontal disease, the chief cause of adult tooth loss. The etiology research involves studies in basic immunology, collagen chemistry, and cell biology.

In the Developmental Biology and Oral-Facial Anomalies Program, support is focused on normal and abnormal oral-facial growth, development, and function. Research, which includes work with primates, concentrates on facial deformities, especially cleft lip/cleft palate, malocclusion, and temporomandibular joint disturbances. The NIDR Biomaterials program concentrates on development and testing of improved and new materials for preventive, restorative, and prosthetic dentistry. Materials scientists, toxicologists, and dentists collaborate in the search for new ways to replace teeth. Artificial implants as well as transplants are receiving much attention. Nonhuman primates offer excellent models for such studies.

The Soft Tissue Stomatology Program, a fifth area in the Extramural Programs, supports investigations into the diagnosis, treatment, and prevention of oral cancer, salivary gland disorders, oral-facial ulcerations, and dental pulp disorders. The newest area, the General Oral Sciences Program, supports basic biologic and behavioral sciences research relating to dentistry. Dental pain control and anaesthesiology are prime concerns of this program.

Use of Nonhuman Primates

Investigations in all six NIDR programs involve a variety of experimental animals, including nonhuman primates. The use of primates in biomedical research has increased rapidly during the past decade; last year, the NIH commitment to studies using primates approximated $ 3,000,000.00 for a broad range of dental research.

Studies of particular interest to this section[2] of the Congress Craniofacial Biology Program, that is, investigations in the developmental and functional aspects of the craniofacial complex, are generally assigned to the NIDR's Developmental Biology and Oral-Facial Anomalies Program. This program has as a prime concern the quest for ways to help children born with cleft lip and/or cleft palate and ways to prevent the birth defects which every year mar one of every 600 to 650 infants. The etiology of this malformation is being studied extensively since it can be produced experimentally in rodents and, to a limited extent, in monkeys. Spontaneous cases of cleft lip/cleft palate also occur in nonhuman primates. A few years ago, a male rhesus macaque with cleft lip and cleft palate was born at the Hazelton Laboratories in Virginia to a mother with no drug history. When we learned of the animal, we were able to invite the attention of Dr. ORVILLE SMITH, Director of the Washington Regional Primate Research Center in Seattle, to the case. Dr. SMITH's group is rearing this monkey at the Washington Center. The animal is very healthy and can manage the eating of regular foods without the benefits of surgery or prosthetic devices.

2   The Symposium on Developmental and Functional Anatomy and the Discussion Seminar on Application to Clinical Problems of Basic Research in Craniofacial Biology. [Ed.]

One scientist supported by the NIDR, Dr. GEORGE J. CHIERICI of the University of California in San Francisco, is trying to determine whether associated abnormalities found in cleft palate children are primary defects due to inadequate cell potential or whether the defects are secondary adaptations to the cleft. To answer this question, Dr. CHIERICI surgically creates a cleft palate in normal monkeys and then carefully follows their facial growth and development. He finds that animals with experimentally-created clefts develop abnormalities similar to those seen in children with clefts. This finding suggests that the abnormalities are secondary adaptations to the cleft and that in treating children efforts should be made to prevent these secondary adaptations or to utilize them to advantage.

Another investigator at the University of California in San Francisco, Dr. EGIL HARVOLD, is using the rhesus macaque to study the influence of jaw position on occlusion. By lowering the position of the animal's mandible, Dr. HARVOLD has created malocclusion.

One of the most fascinating studies we support, philosophically speaking, concerns basic speech mechanisms. In addition to studying the brain's encoding and decoding abilities, Dr. PHILIP LIEBERMAN of the Haskins Laboratories and the University of Connecticut in Storrs, is examining the throat structure and its role in speech. He concludes that the shape of the throat determines whether or not an animal produces speech sounds. He found that the monkey's throat shape makes it impossible for that animal to produce the many sounds of human speech. Similarly, their immature throat shape prevents human infants from producing mature speech.

State-of-the-Art Workshops

In addition to supporting research and research training, the NIDR has sponsored a series of state-of-the-art workshops. The purpose of these workshops is to assess the status of research in a given field and to point out promising areas for further study. They are structured so that small groups can work together intensively and freely assess progress in their field. Their evaluations are published and are a useful resource for investigators in the field and for the NIDR staff in program planning.

One of our state-of-the-art evaluations concerned nutrition in nonhuman primates. Held in 1969, that workshop was sponsored jointly by the NIDR and the Primate Research Centers Section of the NIH Division of Research Resources. Because standards of primate nutrition are not well established and because diets and nutritional status of experimental animals are basic to any research project, the consultants focused their attention on optimum diets for maintenance and growth of various primate species. The discussions and recommendations on standardized diets for primate species were published by Academic Press in 1970 in a volume entitled *Feeding and Nutrition of Non-Human Primates,* edited by ROBERT S. HARRIS (then at Massachusetts Institute of Technology). I should like to add that Professor SHERWOOD L. WASHBURN, to whom the Fourth International Congress of Primatology is dedicated, participated in the workshop on the nutrition of nonhuman primates.

Projections – NIDR and Future Support

I would like to conclude my remarks by listing some of the areas of research to which we feel primate research can contribute greatly. There is a need for establishing a good primate model for dental caries. Although the English are working with primates, little of this work has been done in this country. Similarly, more primate research in periodontal disease is needed. Primate studies of oral muscle forces, oral habits, and tissue responses to various mechanical and physical forces encountered in orthodontic therapy also are lacking, as are studies on craniofacial growth and development and on tooth eruption. Interest is also high in the area of artificial tooth implants. The NIDR describes its areas of interest in a new publication entitled 'Oral Disease – Target for the 70's: A Five-Year Plan of the NIDR for Optimum Development of the Nation's Dental Research Effort'.

Author's address: Dr. K. KENNETH HISAOKA, National Institute of Dental Research-EP, NIH, Westwood Bldg., Room 503, *Bethesda, MD 20014* (USA)

Subject Index